PE EXAM PRE

CIVIL ENGINEERING
PE PROBLEMS & SOLUTIONS

Seventeenth Edition

James H. Banks, PhD, San Diego State University

Braja M. Das, PhD, PE, California State University, Sacramento

Bruce E. Larock, PhD, PE, University of California, Davis

Thomas B. Nelson, PhD, PE, University of Wisconsin—Platteville

Donald G. Newnan, PhD, PE, San Jose State University

Robert W. Stokes, PhD, Kansas State University

Alan Williams, PhD, SE, Chartered Eng., California Department of Transportation

Kenneth J. Williamson, PhD, PE, Oregon State University

KAPLAN) AEC EDUCATION

This publication is designed to provide accurate and authoritative information in regard to the subject matter covered. It is sold with the understanding that the publisher is not engaged in rendering legal, accounting, or other professional service. If legal advice or other expert assistance is required, the services of a competent professional person should be sought.

Vice President & General Manager: David Dufresne
Vice President of Product Development & Publishing: Evan M. Butterfield
Editorial Project Manager: Laurie McGuire
Director of Production: Daniel Frey
Production Editor: Samantha Raue
Production Artist: Caitlin Ostrow
Creative Director: Lucy Jenkins

© 2008 by Dearborn Financial Publishing, Inc.®

Published by Kaplan® AEC Education
30 South Wacker Drive
Chicago, IL 60606-7481
(312) 836-4400
www.kaplanaecengineering.com

All rights reserved. The text of this publication, or any part thereof, may not be reproduced in any manner whatsoever without written permission in writing from the publisher.

Printed in the United States of America.

08   10   9   8   7   6   5   4   3   2

CONTENTS

CHAPTER 1 — **Introduction** 1
HOW TO USE THIS BOOK 1
BECOMING A PROFESSIONAL ENGINEER 2
CIVIL ENGINEERING PROFESSIONAL ENGINEER EXAM 2
ACKNOWLEDGMENTS 7
ERRATA 7

CHAPTER 2 — **Building Structures** 9
PROBLEMS 9
SOLUTIONS 17
REFERENCES 28

CHAPTER 3 — **Bridge Structures** 29
PROBLEMS 29
SOLUTIONS 38
REFERENCES 55

CHAPTER 4 — **Foundations and Retaining Structures** 57
PROBLEMS 57
SOLUTIONS 64
REFERENCES 74

CHAPTER 5 — **Hydraulics** 75
PROBLEMS 75
SOLUTIONS 84
REFERENCES 102

CHAPTER 6 — **Engineering Hydrology** 103
PROBLEMS 103
SOLUTIONS 109
REFERENCES 120

CHAPTER 7 — Water Quality, Treatment, and Distribution 121
PROBLEMS 121
SOLUTIONS 126
REFERENCES 129

CHAPTER 8 — Wastewater Treatment 131
PROBLEMS 131
SOLUTIONS 135
REFERENCES 140

CHAPTER 9 — Geotechnical Engineering 141
PROBLEMS 141
SOLUTIONS 149
REFERENCES 158
RECOMMENDATIONS FOR FURTHER STUDY 159

CHAPTER 10 — Transportation Engineering 161
PROBLEMS 161
SOLUTIONS 169
REFERENCES 178
RECOMMENDED EXAM REFERENCES 179

CHAPTER 11 — Construction Engineering 181
PROBLEMS 181
SOLUTIONS 184
REFERENCES 187

APPENDIX A — Engineering Economics 189
PROBLEMS 189
SOLUTIONS 200

CHAPTER 1

Introduction

OUTLINE

HOW TO USE THIS BOOK 1

BECOMING A PROFESSIONAL ENGINEER 2
Education ■ Fundamentals of Engineering (FE/EIT) Exam ■ Experience ■ Professional Engineer Exam

CIVIL ENGINEERING PROFESSIONAL ENGINEER EXAM 2
Examination Development ■ Examination Structure ■ Design Standards ■ Exam Dates ■ Exam Procedure ■ Preparing for and Taking the Exam ■ Exam Day Preparations ■ What to Take to the Exam ■ Examination Scoring and Results

ACKNOWLEDGMENTS 7

ERRATA 7

HOW TO USE THIS BOOK

Civil Engineering PE Problems & Solutions and its companion texts form a three-step approach to preparing for the Principles and Practice of Engineering (PE) exam:

- *Civil Engineering PE License Review* contains the conceptual review of civil engineering topics for the exam, including key terms, equations, solved examples, analytical methods, and reference data.

- *Civil Engineering PE Problems & Solutions* provides problems for you to solve to test your understanding of concepts and techniques. Ideally, you should solve these problems after completing your conceptual review. Then, compare your solutions to the detailed solutions provided to get a sense of how well you have mastered the content and determine which topics you need to review further.

- *Civil Engineering PE Sample Exam* provides complete morning and afternoon exam sections so that you can simulate the experience of taking the PE exam within its actual time constraints and with questions that match the exam format. After you are satisfied with your review of concepts and problem-solving techniques, take the sample exam to ensure your readiness for the real exam.

BECOMING A PROFESSIONAL ENGINEER

To achieve registration as a professional engineer there are four distinct steps: (1) education, (2) the Fundamentals of Engineering/Engineer-In-Training (FE/EIT) exam, (3) professional experience, and (4) the Professional Engineer (PE) exam, more formally known as the Principles and Practice of Engineering Exam. These steps are described in the following sections.

Education

The obvious appropriate education is a BS degree in civil engineering from an accredited college or university. This is not an absolute requirement. Alternative, but less acceptable, education is a BS degree in something other than civil engineering, or a degree from a nonaccredited institution, or four years of education but no degree.

Fundamentals of Engineering (FE/EIT) Exam

Most people are required to take and pass this eight-hour multiple-choice examination. Different states call it by different names (Fundamentals of Engineering, EIT, or Intern Engineer), but the exam is the same in all states. It is prepared and graded by the National Council of Examiners for Engineering and Surveying (NCEES). Review materials for this exam are found in other Kaplan® AEC Education books, such as *Fundamentals of Engineering FE/EIT Exam Preparation*.

Experience

Typically one must have four years of acceptable experience before being permitted to take the Professional Engineer exam. Both the length and character of the experience will be examined. It may, of course, take more than four years to acquire four years of acceptable experience.

Professional Engineer Exam

The second national exam is called Principles and Practice of Engineering by NCEES, but most people call it the Professional Engineer or PE exam. All states, plus Guam, the District of Columbia, and Puerto Rico use the same NCEES exam.

CIVIL ENGINEERING PROFESSIONAL ENGINEER EXAM

The reason for passing laws regulating the practice of civil engineering is to protect the public from incompetent practitioners. Beginning about 1907, the individual states began passing *title* acts regulating who could call themselves a civil engineer. As the laws were strengthened, the *practice* of civil engineering was limited to those who were registered civil engineers or working under the supervision of a registered civil engineer. There is no national registration law; registration is based on individual state laws and is administered by boards of registration in each of the states. You can find a list of contact information for and links to the various state boards of registration at the Kaplan AEC Web site: *www.kaplanaecengineering.com*. This list also shows the exam registration deadline for each state.

Examination Development

Initially, the states wrote their own examinations, but beginning in 1966, the NCEES took over the task for some states. Now the NCEES exams are used by all states. This greatly eases the ability of a civil engineer to move from one state to another and achieve registration in the new state.

The development of the civil engineering exam is the responsibility of the NCEES Committee on Examinations for Professional Engineers. The committee is composed of people from industry, consulting, and education plus consultants and subject-matter experts. The starting point for the exam is a civil engineering task analysis survey, which the NCEES does at roughly five- to ten-year intervals. People in industry, consulting, and education are surveyed to determine what civil engineers do and what knowledge is needed. From this, NCEES develops what it calls a "matrix of knowledge," which forms the basis for the civil engineering exam structure described in the next section.

The actual exam questions are prepared by the NCEES committee members, subject matter experts, and other volunteers. All people participating must hold professional registration. Using workshop meetings and correspondence by mail, the questions are written and circulated for review. Although based on an understanding of engineering fundamentals, the problems require the application of practical professional judgement and insight.

Examination Structure

The exam is organized into breadth and depth sections.

The morning breadth exam consists of 40 multiple-choice questions covering the following areas of civil engineering: water resources and environmental, geotechnical, structural, transportation, and construction. Each topic area represents 20 percent of the exam questions. You will have four hours to complete the breadth exam.

The afternoon depth portion is actually five exams—one on each of the morning breadth topics. You can choose the depth exam you wish to take; the obvious choice is whichever one best matches your training and professional practice. You will have four hours to answer the 40 multiple-choice questions that make up the depth exam.

Both the breadth and depth questions include four possible answers (A, B, C, D) and are objectively scored by computer.

For more information on the topics, subtopics, and their relative weights on the breadth and depth portions, visit the NCEES Web site at *www.ncees.org*.

Design Standards

The PE exam for civil engineering incorporates design standards and codes from several widely used industry references. Candidates can expect to see structural, transportation, and construction engineering questions that require them to apply equations, safety factors, and other criteria from the design standards.

NCEES posts the relevant standards for each exam on its Web site. You should download these lists and become familiar with the organization and application of each reference. You probably will want to bring a copy of relevant sections of each reference to the exam itself.

Exam Dates

The National Council of Examiners for Engineering and Surveying (NCEES) prepares Civil Engineering Professional Engineer exams for use on a Friday in April and October each year. Some state boards administer the exam twice a year, whereas others offer the exam once a year. The scheduled exam dates for the next ten years can be found on the NCEES Web site: *www.ncees.org/exams/schedules/*.

People seeking to take a particular exam must apply to their state board several months in advance.

Exam Procedure

Before the four-hour morning session begins, proctors will pass out an exam booklet, answer sheet, and mechanical pencil to each examinee. The provided pencil is the only writing instrument you are permitted to use during the exam. If you need an additional pencil during the exam, a proctor will supply one.

Fill in the answer bubbles neatly and completely. Questions with two or more bubbles filled in will be marked as incorrect, so if you decide to change an answer, be sure to erase your original answer completely.

The afternoon session will begin following a one-hour lunch break.

In both the morning and afternoon sessions, if you finish more than 15 minutes early, you may turn in your booklet and answer sheet and leave. In the last 15 minutes, however, you must remain to the end of the exam to ensure a quiet environment for those still working and an orderly collection of materials.

Preparing for and Taking the Exam

Give yourself time to prepare for the exam in a calm and unhurried way. Many candidates like to begin several months before the exam. Target a number of hours per day or week that you will study and reserve blocks of time for doing so. Creating a review schedule on a topic-by-topic basis is a good idea. Remember to allow time for both reviewing concepts and solving practice problems. You may want to prioritize the time you spend reviewing specific topics according to their relative weight on the exam, as identified by NCEES, or by your areas of strength and weakness.

In addition to review work that you do on your own, you may want to join a study group or take a review course. A group study environment might help you stay committed to a study plan and schedule. Group members can create additional practice problems for one another and share tips and tricks.

People familiar with the psychology of exam taking have several suggestions for people as they prepare to take an exam:

- Exam taking involves, really, two skills. One is the skill of illustrating knowledge that you know. The other is the skill of exam taking. The first may be enhanced by a systematic review of the technical material. Exam-taking skills, on the other hand, may be improved by practice with similar problems presented in the exam format.

- Since there is no deduction for guessing on the multiple-choice problems, an answer should be given for all of them. Even when one is going to guess, a logical approach is to attempt to eliminate one or two of the four alternatives. If this can be done, the chance of selecting a correct answer obviously improves from 1 in 4 to 1 in 3 or 1 in 2.

- Plan ahead with a strategy. Which is your strongest area? Can you expect to see several problems in this area? What about your second strongest area? What is your weakest?

- Plan your time allocation ahead. Compute how much time you will allow for each of the five subject areas in the breadth exam and the relevant topics in the depth exam. You might allocate a little less time per problem for those areas in which you are most proficient, leaving a little more time in subjects that are more difficult for you. Your time plan should include a reserve block for especially difficult problems, for checking your scoring sheet, and to make last-minute guesses on problems you did not work. Your strategy might also include time allotments for two passes through the exam—the first to work all problems for which answers are obvious to you, the second to return to the more complex, time-consuming problems and the ones at which you might need to guess. A time plan gives you the confidence of being in control and keeps you from making the serious mistake of misallocating time.

- Read all four multiple-choice answers before making a selection. An answer in a multiple-choice question is sometimes a plausible decoy—not the best answer.

- Do not change an answer unless you are absolutely certain you have made a mistake. Your first reaction is likely to be correct.

- Do not sit next to a friend, a window, or other potential distractions.

Exam Day Preparations

The exam day will be a stressful and tiring one. This will be no day to have unpleasant surprises. For this reason, we suggest making an advance visit to the examination site. Try to determine such items as the following:

- How much time should I allow for travel to the exam on that day? Plan to arrive about 15 minutes early. That way you will have ample time but not too much time. Arriving too early and mingling with others who are also anxious will increase your anxiety and nervousness.

- Where will I park?

- How does the exam site look? Will I have ample workspace? Where will I stack my reference materials? Will it be overly bright (sunglasses), cold (sweater), or noisy (earplugs)? Would a cushion make the chair more comfortable?

- Where are the drinking fountain and lavatory facilities?

- What about food? Should I take something along for energy in the exam? A bag lunch during the break probably makes sense.

What to Take to the Exam

The NCEES guidelines say you may bring only the following reference materials and aids into the examination room for your personal use:

- Handbooks and textbooks, including the applicable design standards.

- Bound reference materials, provided the materials remain bound during the entire examination. The NCEES defines *bound* as books or materials fastened securely in their covers by fasteners that penetrate all papers. Examples are ring binders, spiral binders and notebooks, plastic snap binders, brads, screw posts, and so on.

- A battery-operated, silent, nonprinting, noncommunicating calculator from the NCEES list of approved calculators. For the most current list, see the NCEES Web site (*www.ncees.org*). You also need to determine whether or not your state permits preprogrammed calculators. Bring extra batteries for your calculator just in case; many people feel that bringing a second calculator is also a very good idea.

At one time NCEES had a rule barring "review publications directed principally toward sample questions and their solutions" in the exam room. This set the stage for restricting some kinds of publications from the exam. *State boards may adopt the NCEES guidelines or adopt either more or less restrictive rules.* Thus an important step in preparing for the exam is to know what will—and will not—be permitted. We suggest that, if possible, you obtain a written copy of your state's policy for the specific exam you will be taking. Occasionally there has been confusion at individual examination sites, so a copy of the exact applicable policy will not only allow you to prepare your materials carefully and correctly, but will also ensure that the exam proctors will allow all proper materials that you bring to the exam.

As a general rule, we recommend that you plan well in advance what books and materials you want to take to the exam. Then they should be obtained promptly so you use the same materials in your review that you will have in the exam.

License Review Books

The review books you use to prepare for the exam are good choices to bring to the exam itself. After weeks or months of studying, you will be very familiar with their organization and content, so you'll be able to locate quickly the material you want to reference during the exam. Keep in mind the caveat just discussed—some state boards will not permit you to bring in review books that consist largely of sample questions and answers.

Textbooks

If you still have your university textbooks, they are the ones you should use in the exam, unless they are too out-of-date. To a great extent, the books will be like old friends, with familiar notation.

Bound Reference Materials

The NCEES guidelines suggest that you can take any reference materials you wish, so long as you prepare them properly. You could, for example, prepare several volumes of bound reference materials with each volume intended to cover a particular category of problem. Maybe the most efficient way to use this book would be to cut it up and insert portions of it in your individually prepared bound materials. Use tabs so that specific material can be located quickly. If you do a careful and systematic review of civil engineering and prepare a lot of well-organized materials, you just may find that you are so well prepared that you will not have left anything of value at home.

Other Items

In addition to the reference materials just mentioned, you should consider bringing the following to the exam:

- *Clock*—You must have a time plan and a clock or wristwatch.

- *Exam Assignment Paperwork*—Take along the letter assigning you to the exam at the specified location. To prove you are the correct person, also bring identification with your name and picture.

- *Items Suggested by Advance Visit*—If you visit the exam site, you probably will think of an item or two that you need to add to your list.

- *Clothes*—Plan to wear comfortable clothes. You probably will do better if you are slightly cool.

- *Box for Everything*—You need to be able to carry all your materials to the exam and have them conveniently organized at your side. Probably a cardboard box is the answer.

Examination Scoring and Results

The questions are machine-scored by scanning. The answer sheets are checked for errors by computer. Marking two answers to a question, for example, will be detected and no credit will be given.

Your state board will notify you whether you have passed or failed roughly three months after the exam. Candidates who do not pass the exam the first time may take it again. If you do not pass, you will receive a report listing the percentages of questions you answered correctly for each topic area. This information can help focus the review efforts of candidates who need to retake the exam.

The PE exam is challenging, but analysis of previous pass rates shows that the majority of candidates do pass it the first time. By reviewing appropriate concepts and practicing with exam-style problems, you can be in that majority. Good luck!

ACKNOWLEDGMENTS

For the review of new construction engineering problems in this edition, the publisher is grateful to Denise D. Gravitt, PhD, CIT, CDS, Western Kentucky University.

ERRATA

The author and publisher of this book have been careful to avoid errors, employing technical reviewers, copy editors, and proofreaders to ensure the material is as flawless as possible. Any known errata and corrections are posted on the product page at our Web site, *www.kaplanAECengineering.com*. If you believe you have discovered an inaccuracy, please notify the engineering editor at Kaplan AEC Education:

E-mail: engineeringpress@kaplan.com
Fax: 312-836-9958
Mail: Kaplan AEC Education
 30 S. Wacker Drive, Suite 2500
 Chicago, IL 60606

CHAPTER 2

Building Structures

PROBLEMS

2.1 A 20-inch-diameter, spirally reinforced column with 1½-inch cover to the spiral is reinforced with ten No. 9 bars. The concrete strength is 5000 pounds per square inch, and the tensile strength of the reinforcement is 60,000 pounds per square inch. The unsupported length of the column is 9 feet, and it is braced against side sway.

a. What is the effective length factor?
   a. 0.65
   b. 0.80
   c. 1.00
   d. 1.20

b. The column slenderness ratio is most nearly:
   a. 19.4
   b. 21.0
   c. 21.6
   d. 22.0

c. The column supports only axial load, and the design axial load strength is given most nearly by:
   (Do not use tables or charts)
   a. 600 kips
   b. 800 kips
   c. 1000 kips
   d. 1100 kips

d. The minimum permitted diameter of the spiral reinforcement is:
   a. ¼ inch
   b. ⅜ inch
   c. ½ inch
   d. ⅝ inch

e. The maximum allowable pitch of the spiral is most nearly:
   a. 1.25 inches
   b. 1.50 inches
   c. 1.75 inches
   d. 2.00 inches

Assuming factored moments from gravity loads of $M_{1b} = M_{2b} = 250$ kip-feet are applied at the ends of the column, which is 12 feet long, is bent in single curvature, and has a factored axial load of 700 kips and a value for the critical load of $P_c = 4000$ kips, then:

f. The moment magnification factor is most nearly:
   a. 1.0
   b. 1.2
   c. 1.3
   d. 1.4

2.2 A W30 × 116 beam with f_y = 36 ksi has the top flange restrained against lateral displacement and rotation and has an unstiffened web. A point load is applied to the top flange at 2 feet from one support. The beam spans 40 feet and $1.5M_a < M_y$.

a. The maximum value of the point load that may be applied without web yielding occurring is most nearly:
   a. 89 kips
   b. 99 kips
   c. 109 kips
   d. 119 kips

b. The maximum value of the point load that may be applied without web crippling occurring is most nearly:
   a. 148 kips
   b. 158 kips
   c. 168 kips
   d. 178 kips

c. The maximum value of the point load that may be applied without side sway web buckling occurring is most nearly:
   a. 148 kips
   b. 158 kips
   c. 168 kips
   d. 178 kips

Lateral restraint to both flanges and a pair of full-depth stiffeners are provided at the location of the point load. The stiffeners are ¾ × 4½ inch with 1-inch corner snips and ⁵⁄₁₆ fillet welds.

d. The load producing the maximum allowable bearing stress on the stiffener is most nearly:
   a. 155 kips
   b. 160 kips
   c. 165 kips
   d. 170 kips

e. The maximum capacity of the stiffener-to-web weld is most nearly:
   a. 468 kips
   b. 478 kips
   c. 488 kips
   d. 498 kips

f. The length of web contributing to the equivalent load-carrying column is most nearly:
   a. 11 inches
   b. 12 inches
   c. 13 inches
   d. 14 inches

g. The slenderness ratio of the equivalent load-carrying column is most nearly:
   a. 11
   b. 11.5
   c. 12
   d. 12.5

h. The maximum capacity of the equivalent load-carrying column is most nearly:
   a. 290 kips
   b. 300 kips
   c. 310 kips
   d. 315 kips

2.3 All bolts in the wind bracing connection shown in Exhibit 2.3 are 1-inch A325 slip-critical bolts in standard holes. All structural sections are Grade A36 steel. The effects of prying action may be neglected.

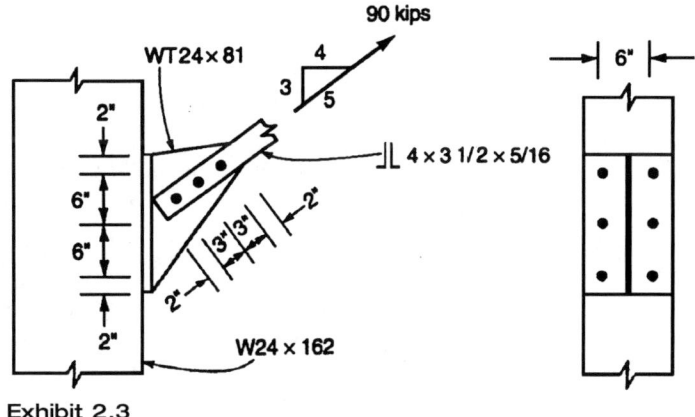

Exhibit 2.3

a. The total shear force applied to the connection at the column flange is most nearly:
   a. 45 kips
   b. 54 kips
   c. 63 kips
   d. 72 kips

b. The total tensile force applied to the connection at the column flange is most nearly:
   a. 45 kips
   b. 54 kips
   c. 63 kips
   d. 72 kips

c. The allowable shear capacity of the connection at the column flange is most nearly:
   a. 46 kips
   b. 51 kips
   c. 56 kips
   d. 61 kips

d. The allowable tensile capacity of the connection at the column flange is most nearly:
   a. 181 kips
   b. 191 kips
   c. 201 kips
   d. 211 kips

e. The allowable bearing capacity of the WT flange is most nearly:
   a. 360 kips
   b. 365 kips
   c. 370 kips
   d. 375 kips

f. The allowable shear capacity of the bolts connecting the angle brace to the tee section is most nearly:
   a. 71 kips
   b. 76 kips
   c. 81 kips
   d. 86 kips

g. The allowable bearing capacity of the angle brace is most nearly:
   a. 80 kips
   b. 85 kips
   c. 90 kips
   d. 95 kips

h. The capacity of the angle brace based on its gross area is most nearly:
   a. 92 kips   c. 102 kips
   b. 97 kips   d. 107 kips

i. The effective net area of the angle brace is most nearly:
   a. 2.0 square inches   c. 3.0 square inches
   b. 2.5 square inches   d. 3.5 square inches

j. The capacity of the angle brace based on its effective net area is most nearly:
   a. 75 kips   c. 85 kips
   b. 80 kips   d. 90 kips

2.4 Details of a single-story wood-framed building are shown in Exhibit 2.4. The governing lateral wind load is indicated.

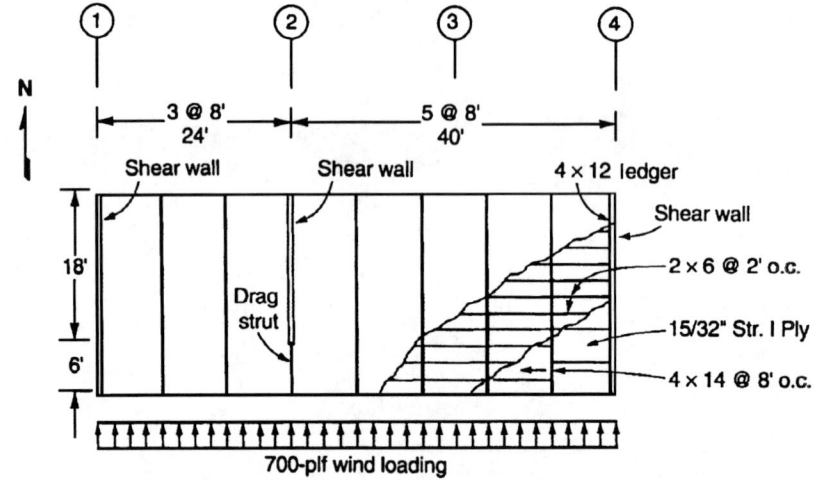

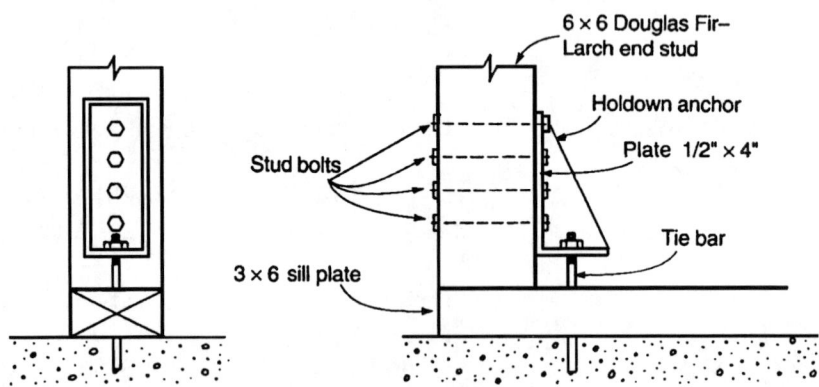

Exhibit 2.4

a. The unit shear at grid line 4 is most nearly:
   a. 490 pounds per linear foot
   b. 520 pounds per linear foot
   c. 550 pounds per linear foot
   d. 580 pounds per linear foot

b. The required spacing of 10-penny nails along grid line 4 is most nearly:
   a. 2 inches   c. 3 inches
   b. 2 ½ inches   d. 4 inches

c. The required spacing of 10-penny nails in the 2 × 6 subpurlins adjacent to grid line 4 is most nearly:
   a. 2 ½ inches   c. 4 inches
   b. 3 inches   d. 6 inches

d. The chord force on the north wall at grid line 3 in most nearly:
   a. 5830 pounds   c. 5930 pounds
   b. 5880 pounds   d. 5980 pounds

e. A lapped splice is provided in the doubled 2 × 6 Douglas Fir–Larch top chord of the north wall at grid line 3. The number of 16-penny common nails required at the splice is most nearly:
   a. 26   c. 32
   b. 29   d. 35

f. The maximum unit shear applied to the shear wall at grid line 2 is most nearly:
   a. 1120 pounds per linear foot
   b. 1160 pounds per linear foot
   c. 1200 pounds per linear foot
   d. 1240 pounds per linear foot

g. The shear wall at grid line 2 is constructed with 3 × 6 studs at 16 inches on center faced on both sides with ⅜-inch plywood siding. The required spacing of 8-penny nails at the plywood panel edges is most nearly:
   a. 2 inches   c. 4 inches
   b. 3 inches   d. 6 inches

h. The height of the shear wall on grid line 2 is 12 feet. Assuming that the end stud, the holdown anchor, and the bolt spacing shown in the detail are adequate, the diameter of the four bolts into the stud is:
   a. ¾ inch   c. 1 inch
   b. ⅞ inch   d. 1 ⅛ inch

i. What is the required diameter of the tie bar?
   a. ¾ inch   c. 1 inch
   b. ⅞ inch   d. 1 ⅛ inch

j. Anchor bolts are provided in the sill plate at a spacing of 16 inches. What diameter is required for these anchor bolts?
   a. ½ inch   c. ¾ inch
   b. ⅝ inch   d. ⅞ inch

2.5 Exhibit 2.5 shows a solid grouted 8-inch concrete block masonry beam simply supported over an effective span of 20 feet for both vertical and lateral loads. The vertical load (including self-weight of the beam) is 1500 pounds per foot. The horizontal load (due to wind) is 70 pounds per foot. The beam has a masonry strength of 1500 pounds per square inch and a modulus of elasticity of 1000 kips per square inch and is reinforced with Grade 60 bars.

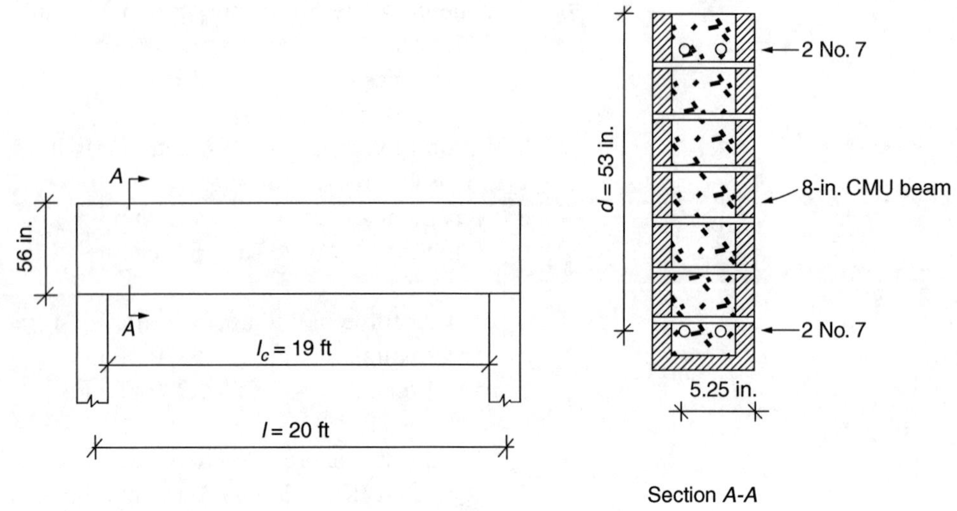

14 Chapter 2 Building Structures

Exhibit 2.5

a. The bending moment due to vertical loads is most nearly:
   a. 70 kip-feet
   c. 80 kip-feet
   b. 75 kip-feet
   d. 85 kip-feet

b. The reinforcement ratio ρ for vertical loads is most nearly:
   a. 0.0029
   c. 0.0039
   b. 0.0034
   d. 0.0044

c. The reinforcement stress due to vertical loads is most nearly:
   a. 15,800 pounds per square inch
   b. 15,850 pounds per square inch
   c. 15,900 pounds per square inch
   d. 15,950 pounds per square inch

d. The bending moment due to lateral loads is most nearly:
   a. 3.0 kip-feet
   c. 4.0 kip-feet
   b. 3.5 kip-feet
   d. 4.5 kip-feet

e. The reinforcement ratio ρ for lateral loads is most nearly:
   a. 0.0030
   c. 0.0040
   b. 0.0035
   d. 0.0045

f. The reinforcement stress due to lateral loads is most nearly:
   a. 7500 pounds per square inch
   b. 7600 pounds per square inch
   c. 7700 pounds per square inch
   d. 7800 pounds per square inch

g. The total stress in the reinforcement due to vertical and lateral loads is most nearly:
   a. 23,300 pounds per square inch
   b. 23,400 pounds per square inch
   c. 23,500 pounds per square inch
   d. 23,600 pounds per square inch

h. The allowable reinforcement stress is most nearly:
   a. 20,000 pounds per square inch
   b. 24,000 pounds per square inch
   c. 26,600 pounds per square inch
   d. 32,000 pounds per square inch

i. The shear stress at the critical section due to vertical loads is most nearly:
   a. 17 pounds per square inch
   b. 22 pounds per square inch
   c. 27 pounds per square inch
   d. 32 pounds per square inch

j. The allowable shear stress in the masonry is most nearly:
   a. 24 pounds per square inch
   b. 29 pounds per square inch
   c. 34 pounds per square inch
   d. 39 pounds per square inch

2.6 A typical composite floor in a commercial building consists of a 4-inch concrete slab supported on W18 × 35, Grade 50, beams at 6-foot centers. The beams span 30 feet and are each provided with 24 studs of ⅝-inch diameter × 2½ inches long.

Determine the maximum moment that the composite beams can support.

2.7 A 4 × 4¼ inch plate is connected to a ½-inch gusset plate with E70XX fillet welds. The plates are of A36 grade steel. Determine the size of weld required to support the 30 kips load.

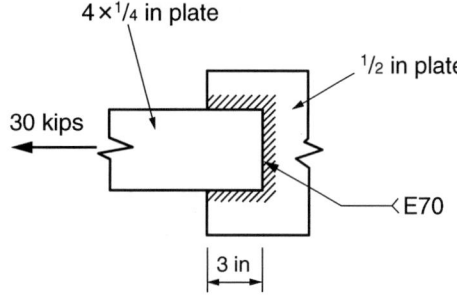

Exhibit 2.7

2.8 The Douglas Fir–Larch column, select structural grade, shown in Exhibit 2.8, supports the roof and floor framing of a two-story commercial building. The lateral wind load of 50 pounds per linear foot causes bending about the strong axis of the post, which is braced about the weak axis at midheight and pinned at both ends. The axial loading consists of 3 kips total dead load, 3 kips roof live load (nonsnow), and 5 kips floor live load.

Determine if the column is adequate.

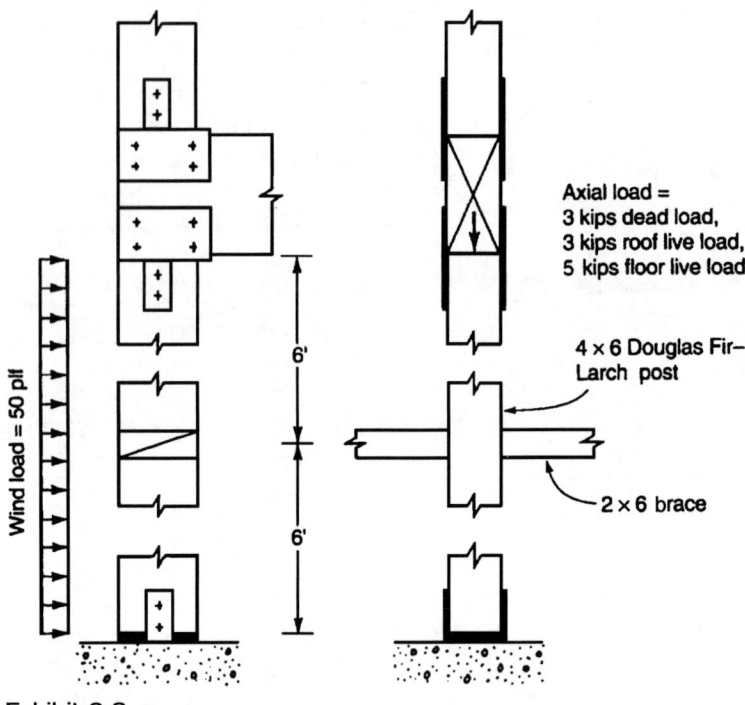

Exhibit 2.8

2.9 The nominal 8-inch solid-grouted, concrete masonry retaining wall shown in Exhibit 2.9 has a specified strength of 1500 pounds per square inch and a modulus of elasticity of 1125 kips per square inch. Reinforcement consists of Number 5 Grade 60 bars at 16-inch centers. The retained soil has an equivalent fluid pressure of 35 pounds per square foot per foot.

Neglecting the self-weight of the wall, determine if the stresses in the reinforcement and the masonry are satisfactory.

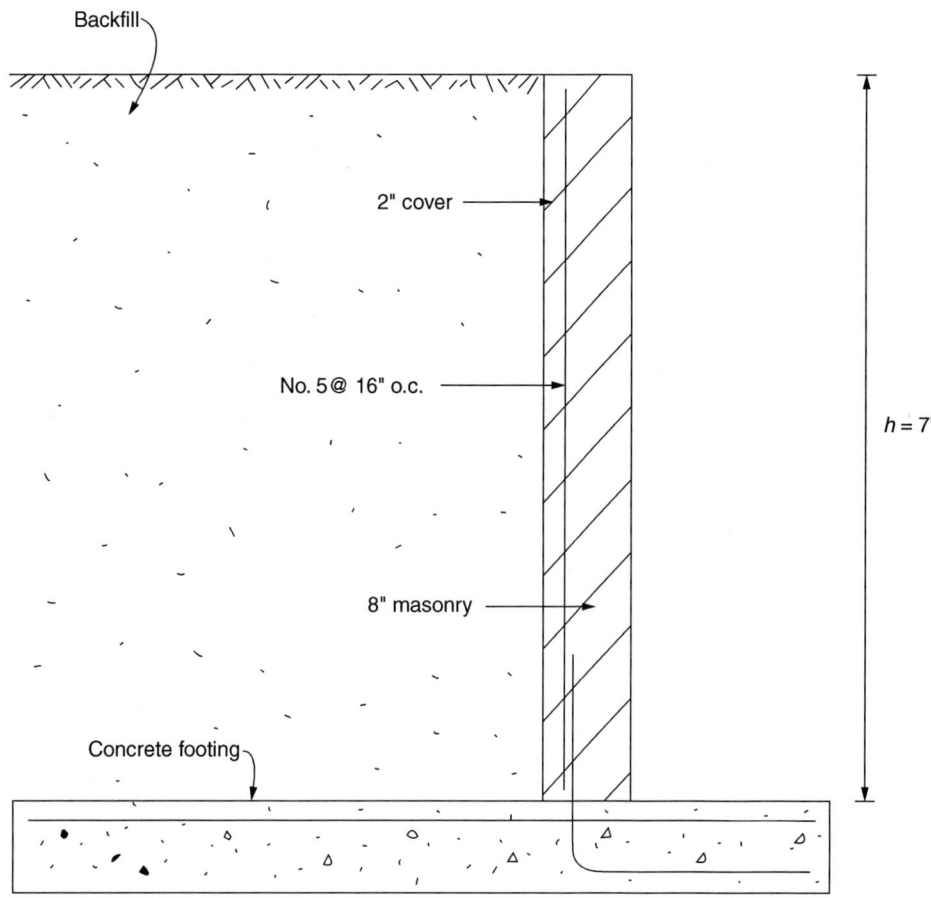

Exhibit 2.9

SOLUTIONS

2.1 a. **c.** From ACI Section 10.12.1, the effective length factor of a column braced against side sway is $k = 1.0$.

b. **c.** The radius of gyration, in accordance with ACI Section 10.11.2, is $r = 0.25D = 0.25 \times 20 = 5$ inches. The slenderness ratio is $kl_u/r = 1 \times 9 \times 12/5 = 21.6$.

c. **d.** For a spirally reinforced column, the design axial load strength is given by ACI Equation (10-1) as $\phi P_n = 0.85\phi[0.85 f'_c (A_g - A_{st}) + f_y A_{st}] = 0.85 \times 0.70 [0.85 \times 5(314.2 - 10) + 60 \times 10] = 1126.24$ kips.

d. **b.** In accordance with ACI Section 7.10.4.2, the minimum diameter of the spiral reinforcement is $\tfrac{3}{8}$ inch.

e. **c.** From ACI Equation (10-5), the minimum spiral reinforcement ratio is $\rho_s = 0.45(A_g/A_{ch} - 1)f'_c/f_y = 0.45(314.2/227.0 - 1)5/60 = 0.0144 = A_s \pi (D_c - D_s)/A_{ch}s = 0.11 \times 3.142(17 - 0.375)/(227 \times s)$. Hence, $s = 1.76$ inches maximum.

f. **c.** From ACI Equation (10-13), the moment factor is given by $C_m = 0.6 + 0.4 M_1/M_2 = 0.6 + 0.4 \times 250/250 = 1.0$. From ACI Equation (10-9), the magnification factor for gravity loads is $\delta_{ns} = C_m/(1 - P_u/0.75\,P_c) = 1/[1 - 700/(0.75 \times 4000)] = 1.3$.

2.2 a. **c.** From AISC Section J10.2 and Equation (J10-2), bearing stiffeners are required when the magnitude of the point load exceeds the value $R = 0.66 F_y t_w (N + 5k) = 0.66 \times 36 \times 0.565\,(0 + 5 \times 1.625) = 109$ kips.

b. **b.** From AISC Equation (J10-4), web crippling will occur when the point load exceeds the value

$$R = 40 t_w^2 [1 + 3(N/d)(t_w/t_f)^{1.5}] \sqrt{E F_y t_f / t_w}$$
$$= 40 \times 0.565^2 \times \sqrt{29{,}000(36) \times 0.85/0.565}$$
$$= 160 \text{ kips}$$

c. **c.** In accordance with AISC Section J10.4, side sway web buckling occurs when the factor $R_8 = h b_f / l t_w < 2.3$; $R_8 = 27 \times 10.5/480 \times 0.565 = 1.05 < 2.3$. From AISC Equation (J10-6) side sway web buckling occurs when the point load exceeds the value $R_n/\Omega = R_7(1 + 0.4 R_8^3)/1.76$:

$$R_7 = 960{,}000\, t_w^3 t_f / h^2$$
$$= 960{,}000 \times 0.565^3 \times 0.85/27^2$$
$$= 202$$
$$R_n/\Omega = 202(1 + 0.4 \times 1.05^3)/1.76$$
$$= 168 \text{ kips}$$

d. **d.** From AISC Equation (J7-1), the load producing the maximum allowable bearing stress on the stiffener is $R = 0.90 F_y A_b = 0.90 \times 36 \times 0.75 \times 2\,(4.5 - 1) = 170$ kips.

e. **c.** The maximum capacity of the stiffener-to-web weld is $R = 4 \times 5 \times 0.928\,(28.3 - 2) = 488$ kips.

f. **d.** The length of web contributing to the equivalent load-carrying column is given in AISC Section K1.8 as $b_w = 25 t_w = 25 \times 0.565 = 14.125$ inches.

g. **a.** The moment of inertia of the equivalent column is $I_e = t_s (2 b_s + t_w)^3/12 = 0.75\,(2 \times 4.5 + 0.565)^3/12 = 54.7$ inches^4. The area of the equivalent column is $A_e = 2 b_s t_s + b_w t_w = 2 \times 4.5 \times 0.75 + 14.125 \times 0.565 = 14.7$ inches^2. The radius of gyration of the equivalent column is $r_e \sqrt{I_e/A_e} = 1.93$ inches. The slenderness ratio of the equivalent column, in accordance with AISC Section J10.8, is given by $Kl/r = 0.75 h/r_e = 0.75 \times 28.3/1.93 = 11$.

h. **d.** The allowable stress is obtained from AISC Table 4-22 as $F_a = 21.4$ kips per square inch. The allowable load is $R = F_a A_e = 21.4 \times 14.7 = 315$ kips.

2.3

a. **b.** The total shear force applied to the connection is $P_v = 90 \times 3/5 = 54$ kips.

b. **d.** The total tensile force applied to the connection is $P_H = 90 \times 4/5 = 72$ kips.

c. **c.** For combined shear and direct tension in a slip-critical connection, the allowable shear capacity in accordance with AISC Section J3.9 is subject to the reduction factor

$$K_s = 1 - 1.5 T_d/D_u T_b N_b$$
$$= 1 - 1.5 \times 72/(1.13 \times 51 \times 6)$$
$$= 0.69$$

The shear capacity of the connection is

$$P_s = 6 \times 0.69 \times 13.4$$
$$= 55.5 \text{ kips}$$

d. **d.** From AISC Table 7-2, the tensile capacity of the connection is $P_t = 6 \times 35.3 = 212$ kips.

e. **d.** The capacity of the column flange in bearing is obtained from AISC Table 7-6 as $P_b = 6 \times 51.1 \times t_f = 6 \times 51.1 \times 1.22 = 374$ kips.

f. **c.** From AISC Table 7-3, the capacity of the bolts in double shear connecting the angle brace to the tee section is $P_s = 3 \times 26.9 = 81$ kips.

g. **d.** The capacity of the angle brace in bearing is obtained from AISC Table 7-6 as $P_b = 3 \times 2 \times 51.1 \times t = 3 \times 2 \times 51.1 \times 0.3125 = 96$ kips.

h. **b.** From AISC Section D2, the strength of the brace based on the gross area is $P_t = 0.6 \times F_y A_g = 0.6 \times 36 \times 2 \times 2.25 = 97$ kips.

i. **c.** From AISC Section D3, the net area of the brace is $A_n = A_g - 2td_h = 2 \times 2.25 - 2 \times 0.3125 (1.0 + 0.125) = 3.80$ square inches. The shear lag factor is

$$U = 1 - \bar{x}/l$$
$$= 1 - 1.17/6$$
$$= 0.81$$

The effective net area of the brace is $A_e = UA_n = 0.81 \times 3.80 = 3.08$ square inches.

j. **d.** The strength of the brace based on the effective net area, in accordance with AISC Section D2, is $P_t = 0.5 F_u A_e = 0.5 \times 58 \times 3.08 = 89$ kips.

2.4

a. **d.** The total wind load on the diaphragm between grid lines 2 and 4 is $V = wL = 700 \times 40 = 28,000$ pounds. The shear force along the diaphragm boundary at grid line 4 is $Q = V/2 = 14,000$ pounds. The unit shear along the diaphragm boundary at grid line 4 is $q = Q/B = 14,000/24 = 583$ pounds per linear foot.

b. **d.** The plywood layout is Case 4 with all edges blocked, and from IBC Table 2306.3.1, 10-penny nails at 4-inch spacing provide a capacity, after allowing for the 40 percent increase for wind design as specified in IBC Section 2306.3.2, of $q' = 425 \times 1.4 = 595$ pounds per linear foot, $> q$.

c. **d.** From IBC Table 2306.3.1, 10-penny nails at 6-inch spacing in the 2×6 subpurlins provide a capacity of $q' = 425 \times 1.4 = 595$ pounds per linear foot, $> q$.

d. **a.** The bending moment in the north wall chord at grid line 3 is $M = VL/8 = 28{,}000 \times 40/8 = 140{,}000$ pound-feet. The corresponding chord force is $T = M/B = 140{,}000/24 = 5833$ pounds.

e. **a.** In accordance with IBC Section 2308.9.2.1, a minimum splice length of 48 inches is required in the top chord members. The load duration factor for wind loading, from NDS Section 2.3.2, is $C_D = 1.6$. The basic allowable lateral load capacity of a 16-penny common nail is given by NDS Table 11N as $Z = 141$ pounds. The modified capacity is $Z' = C_D Z = 1.6 \times 141 = 225$ pounds. The number of nails required is $n = T/Z' = 5833/225 = 26$ nails. Providing two rows of nails at a spacing of 3.5 inches fits within a splice length of 48 inches.

f. **d.** The reactions at the shear wall on grid line 2 are $R_{21} = 700 \times 24/2 = 8400$ pounds and $R_{23} = 700 \times 40/2 = 14{,}000$ pounds. The total unit shear in the shear wall is $q_w = (R_{21} + R_{23})/L_w = (8400 + 14{,}000)/18 = 1244$ pounds per linear foot.

g. **b.** From Footnote d of IBC Table 2306.4.1, the allowable shear values for $3/8$-inch plywood may be increased to the values given for 15/32-inch plywood provided studs are spaced at 16 inches on center. The allowable shear values may be doubled when plywood is applied to both faces of the shear wall. For wind design, the shear capacities may be increased by 40 percent. Hence, from IBC Table 2306.4.1, using 8-penny nails at 3-inch spacing in 3×6 studs provides a shear capacity of $q' = 2 \times 1.4 \times 490 = 1372$ pounds per linear foot $> q_w$.

h. **c.** Neglecting the vertical load on the shear wall, the uplift force on the holdown anchor is $T_w = q_w h = 1244 \times 12 = 14{,}928$ pounds. The gross areas of the end stud and the side plate are $A_m = 30.25$ square inches and $A_s = 2$ square inches. $A_m/A_s = 15.13$. Hence, the group action modification factor for four bolts in a row is obtained from NDS Table 10.3.6C as $C_g = 0.95$. The load duration factor for wind loading, from NDS Section 2.3.2, is $C_D = 1.6$. The reference single shear value for a 1-inch diameter bolt in a $5\frac{1}{2}$-inch Douglas Fir–Larch post with loading parallel to the grain is obtained from NDS Table 11B as $Z = 2860$ pounds. Hence, the capacity of four 1-inch bolts is $T_B = 4 C_g C_D Z = 4 \times 0.95 \times 1.6 \times 2860 = 17{,}388$ pounds, $> T_W$.

i. **c.** From AISC Part 7, Table 7-2, the allowable tensile force on a 1-inch diameter A307 tie bar is $T_T = 17{,}700$ pounds, $> T_W$.

j.   **c.** The basic allowable single-shear value for a ¾-inch diameter anchor bolt in the 2½-inch-wide Douglas Fir–Larch sill plate is obtained, from NDS Table 11E, as $Z = 1540$ pounds. The load duration factor for wind loading, from NDS Section 2.3.2, is $C_D = 1.6$. Hence, the modified capacity of one ¾-inch diameter bolt is $Z' = 1.6 \times 1540 = 2464$ pounds. The shear on the sill plate is $q_w = 1244$ pounds per linear foot $= q_w \times 16/12$ pounds per 16 inches $= 1660$ pounds per 16 inches $< Z'$.

2.5   a.   **b.** The bending moment at midspan due to vertical loads is

$$M = w\ell^2/8$$
$$= 1.5 \times 20^2/8$$
$$= 75 \text{ kip-feet}$$

b.   **a.** The reinforcement ratio for vertical loads is

$$\rho = A_s/b_w d$$
$$= 1.20/(7.63 \times 53)$$
$$= 0.00297$$

c.   **d.** The design parameter is

$$\rho n = 0.00297 \times 29$$
$$= 0.08606$$

The flexural stresses may now be obtained using the elastic design method as

$$k = (2\rho n + (\rho n)^2)^{0.5} - \rho n$$
$$= 0.338$$
$$j = 1 - k/3$$
$$= 0.887$$

The reinforcement stress due to vertical loads is

$$f_s = M/jdA_s$$
$$= 75 \times 12 \times 1000/(0.887 \times 53 \times 1.20)$$
$$= 15{,}950 \text{ pounds per square inch}$$

d.   **b.** The bending moment at midspan due to horizontal loads is

$$M = w\ell^2/8$$
$$= 0.07 \times 20^2/8$$
$$= 3.5 \text{ kip-feet}$$

e.   **c.** The reinforcement ratio for horizontal loads is

$$\rho = A_s/b_w d$$
$$= 1.20/(56 \times 5.25)$$
$$= 0.00408$$

f. **b.** The design parameter is

$$\rho n = 0.00408 \times 29$$
$$= 0.1184$$

Using allowable stress design,

$$k = (2\rho n + (\rho n)^2)^{0.5} - \rho n$$
$$= 0.382$$

$$j = 1 - k/3$$
$$= 0.873$$

$$f_s = M/jdA_s$$
$$= 3.5 \times 12 \times 1000/(0.873 \times 5.25 \times 1.20)$$
$$= 7637 \text{ pounds per square inch}$$

g. **d.** Total stress in the reinforcement due to vertical and lateral loads is

$$f_s = 15{,}950 + 7637$$
$$= 23{,}587 \text{ pounds per square inch}$$

h. **b.** The allowable stress in the reinforcement is given by BCRMS Section 2.3.2.1 as

$$F_s = 24{,}000 \text{ pounds per square inch}$$

i. **c.** The critical section for shear, in accordance with BCRMS Section 2.3.5.5, is a distance of $d/2$ from the face of the support. The shear force, due to vertical loads, at a distance of $d/2$ from each support is given by

$$V = w(\ell_c - d)/2$$
$$= 1.5(19 - 53/12)/2$$
$$= 10.94 \text{ kips}$$

The shear stress at a distance of $d/2$ from each support is given by BCRMS Equation (2-19) as

$$f_v = V/b_w d$$
$$= 10.94 \times 1000/(7.63 \times 53)$$
$$= 27.05 \text{ pounds per square inch}$$

j. **d.** The allowable shear stress without shear reinforcement is given by BCRMS Equation (2-20) as

$$F_v = (f'_m)^{0.5}$$
$$= (1500)^{0.5}$$
$$= 38.7 \text{ pounds per square inch}$$

2.6 The effective width of the concrete slab, in accordance with AISC Section I1, is the lesser of $b = L/4 = 30 \times 12/4 = 90$ inches and $b = s = 6 \times 12 = 72$ inches, the latter of which governs.

The total shear strength provided by 12 stud connectors on each side of the beam center with a diameter of 5/8 inch in normal-weight 3 kips per square inch concrete is

$$\Sigma Q_n = 12 \times 14.6$$
$$= 175 \text{ kips}$$

The depth of the compression stress block in the concrete flange is

$$a = \Sigma Q_n/0.85 f'_c$$
$$= 175/(0.85 \times 3 \times 72)$$
$$= 0.95 \text{ inch}$$

The distance from top of steel beam to concrete flange force is

$$Y_2 = Y_{con} - a/2$$
$$= 4 - 0.95/2$$
$$= 3.53 \text{ inches}$$

From AISC Part 3, Table 3-19, for $Y_2 = 3.53$ inches and $\Sigma Q_n = 175$ kips, the allowable flexural strength is

$$M_n/\Omega_b = 248 \text{ kip-feet}$$

2.7 The minimum size of fillet weld for the ½-inch plate is given by AISC Table J2.4 as

$$w_{min} = \tfrac{3}{16} \text{ inch}$$

The maximum size of fillet weld for the ¼-inch plate is given by AISC Section J2.2(b) as

$$w_{max} = \tfrac{1}{4} - \tfrac{1}{16}$$
$$= \tfrac{3}{16} \text{ inch}$$

Hence, a ³⁄₁₆-inch fillet weld is required.

The total length of longitudinally loaded weld is

$$\ell_{wl} = 2 \times 3$$
$$= 6 \text{ inches}$$

The total length of transversely loaded weld is

$$\ell_{wt} = 4 \text{ inches}$$

The allowable shear capacity of ³⁄₁₆-in fillet weld is

$$q_a = 3 \times 0.928$$
$$= 2.78 \text{ kips/inch}$$

Applying AISC Equation (J2-9a), the allowable strength of the connection is

$$R_n/\Omega = (R_{wl} + R_{wt})/\Omega$$
$$= \ell_{wl} q_a + \ell_{wt} q_a$$
$$= 6 \times 2.78 + 4 \times 2.78$$
$$= 27.8 \text{ kips}$$

Applying AISC Equation (J2-9b), the allowable strength of the connection is

$$R_n/\Omega = (0.85R_{wl} + 1.5R_{wt})/\Omega$$
$$= 0.85\ell_{wl}q_a + 1.5\ell_{wt}q_a$$
$$= 0.85 \times 6 \times 2.78 + 1.5 \times 4 \times 2.78$$
$$= 30.9 \text{ kips ... governs}$$
$$> 27.8 \text{ kips}$$
$$> 30 \text{ kips}$$

Hence, the 3/16-inch fillet weld is satisfactory.

2.8 The applicable load combinations for service load design are given in IBC Section 1605.3.1, and the appropriate load duration factors are given in NDS Section 2.3.2. The combinations are as follows:

(i) From IBC Equation (16-9): Dead load plus floor live load with $C_D = 1.00$

(ii) From IBC Equation (16-11): Dead load plus 0.75 × floor plus 0.75 × roof live load with $C_D = 1.25$

(iii) From IBC Equation (16-13): Dead load plus 0.75 × floor plus 0.75 × roof live load plus 0.75 × wind load with $C_D = 1.6$

The relevant properties of the 4 × 6 Select Structural Douglas Fir–Larch post are $b = 3.5$ inches, $d = 5.5$ inches, $A = 19.25$ inches, $S_x = 17.6$ inches^3, $F_b = 1500$ pounds per square inch, $F_c = 1700$ pounds per square inch, and $E_{min} = 0.69 \times 10^6$ pounds per square inch. The size factor for compression, from NDS SUPP Table 4A, is $C_F = 1.1$. The size factor for bending, from NDS SUPP Table 4A, is $C_F = 1.3$.

The relevant column parameters are: l_e/d = slenderness ratio about the strong axis = $K_e l/d = 1 \times 12 \times 12/5.5 = 26.18 < 50$, which is satisfactory, and l_e/b = slenderness ratio about the weak axis = $K_e l/b = 1 \times 6 \times 12/3.5 = 20.57 < 50$, which is satisfactory. Hence, the governing slenderness ratio is about the strong axis.

For load combination (i), the basic compression design value multiplied by all applicable adjustment factors except C_P is given by NDS Section 3.7.1 as

$$F_c^* = C_D C_F F_c$$
$$= 1.0 \times 1.1 \times 1700$$
$$= 1870 \text{ pounds per square inch}$$

The critical buckling design value is

$$F_{cE} = 0.822 E_{min}/(l_e/d)^2$$
$$= 0.822 \times 0.69 \times 10^6/26.18^2$$
$$= 828 \text{ pounds per square inch}$$

$$F = F_{cE}/F_c^*$$
$$= 828/1870$$
$$= 0.44$$
$$c = \text{column parameter}$$
$$= 0.8 \text{ ... for sawn lumber}$$

The column stability factor is given by NDS Equation (3.7-1) as

$$C_P = (1.0 + F)/2c - \sqrt{[(1.0+F)/2c]^2 - F/c}$$
$$= 1.44/1.6 - \sqrt{(1.44/1.6)^2 - 0.44/0.8}$$
$$= 0.39$$

The allowable compression design value parallel to grain is

$$F'_c = C_P F^*_c$$
$$= 0.39 \times 1870$$
$$= 735 \text{ pounds per square inch}$$

The allowable load on the column is given by

$$P = F'_c A$$
$$= 735 \times 19.25$$
$$= 14{,}150 \text{ pounds}$$

The actual load for load combination (i) is

$$P' = 3000 + 5000$$
$$= 8000 \text{ pounds}$$
$$< P \ldots \text{satisfactory}$$

For load combination (ii),

$$F^*_c = 1.25 \times 1.1 \times 1700$$
$$= 2338 \text{ pounds per square inch.}$$
$$F_{cE} = 828 \text{ pounds per square inch}$$
$$F = 828/2338$$
$$= 0.35$$
$$C_P = 1.35/1.6 - \sqrt{(1.35/1.6)^2 - 0.35/0.8}$$
$$= 0.32$$
$$F'_c = 0.32 \times 2338$$
$$= 748 \text{ pounds per square inch}$$

The allowable load on the column is given by

$$P = 748 \times 19.25$$
$$= 14{,}399 \text{ pounds}$$

The actual load for load combination (ii) is

$$P' = 3000 + (5000 + 3000)0.75$$
$$= 6000 \text{ pounds}$$
$$< P \ldots \text{satisfactory}$$

For load combination (iii),

$$F^*_c = 1.6 \times 1.1 \times 1700$$
$$= 2992 \text{ pounds per square inch}$$
$$F_{cE} = 828 \text{ pounds per square inch}$$
$$F = 828/2992$$
$$= 0.28$$

$$C_P = 1.28/1.6 - \sqrt{1.28/(1.6)^2 - 0.28/0.8}$$
$$= 0.26$$
$$F'_c = 0.26 \times 2992$$
$$= 778 \text{ pounds per square inch}$$

For the uniformly distributed wind load and a l_u/d ratio > 7, the effective length is obtained from Figure 2.43 as

$$l_e = 1.63 l_u + 3d$$
$$= 1.63 \times 6 + 3 \times 5.5/12$$
$$= 11.16 \text{ feet}$$

The slenderness ratio is

$$R_B = \sqrt{l_e d/b^2}$$
$$= \sqrt{12 \times 11.16 \times 5.5/3.5^2}$$
$$= 7.75$$
$$< 50 \ldots \text{satisfactory}$$

The critical buckling design value is given by NDS Section 3.3.3.8 as

$$F_{bE} = 1.2 E'_{min}/R_B^2$$
$$= 1.2 \times 0.69 \times 10^6/7.75^2$$
$$= 13{,}786 \text{ pounds per square inch}$$

The reference flexural design value multiplied by all applicable adjustment factors except C_L is

$$F_b^* = F_b C_D C_F$$
$$= 1500 \times 1.6 \times 1.3$$
$$= 3120 \text{ pounds per square inch}$$
$$F = F_{bE}/F_b^*$$
$$= 13{,}786/3120$$
$$= 4.42$$

The beam stability factor is given by NDS Equation (3.3-6) as

$$C_L = (1.0 + F)/1.9 - \sqrt{((1.0 + F)/1.9)^2 - F/0.95}$$
$$= 0.986$$

The allowable flexural design value is

$$F'_b = C_L F_b^*$$
$$= (0.986) \times 3120$$
$$= 3076$$

The actual load for load combination (iii) is

$$P = 3000 + (5000 + 3000)0.75$$
$$= 6000 \text{ pounds}$$

This produces an axial stress of

$$f_c = 312 \text{ pounds per square inch}$$

The bending moment on the post is

$$M = 50 \times 12^2/8$$
$$= 900 \text{ pound-feet}$$

This produces a bending stress of

$$f_b = 900 \times 12/17.6$$
$$= 614 \text{ pounds per square inch}$$

The moment magnification factor for axial compression and flexure is

$$C_m = 1.0 - f_c/F_{cE}$$
$$= 1 - 312/828$$
$$= 0.623$$

The interaction equation for bending and axial load, as given in NDS Section 3.9.2, is

$$(f_c/F_c')^2 + f_b/F_b' \, C_m \leq 1.0$$

The left-hand side of the expression is

$$(312/748)^2 + 614/(3076 \times 0.623) = 0.49$$
$$< 1.0$$

The column is adequate.

2.9 From BCRMS Section 2.3.2 and 2.3.3, the allowable stresses are

$$F_b = 0.33 f_m'$$
$$= 500 \text{ pounds per square inch}$$
$$F_s = 24{,}000 \text{ pounds per square inch}$$

The maximum bending moment in the wall is given by

$$M = qh^3/6$$
$$= 35 \times 7^3/6$$
$$= 2000 \text{ pound-feet per foot}$$

The design parameters are

$$n = E_s/E_m$$
$$= 29{,}000/1125$$
$$= 25.78$$

$$A_s = 0.235 \text{ square inch per foot}$$
$$d = 7.63 - 2 - 0.63/2$$
$$= 5.32 \text{ inches}$$

$$\rho = 0.235/(12 \times 5.32)$$
$$= 0.00368$$
$$k = (n^2\rho^2 + 2n\rho)^{0.5} - n\rho$$
$$= 0.351$$

$$j = 1 - k/3$$
$$= 0.883$$

$$f_b = 2M/jkbd^2 = 456 \text{ pounds per square inch}$$
$$< 500 \text{ pounds per square inch} \ldots \text{satisfactory}$$

$$f_s = M/jdA_s$$
$$= 21{,}750 \text{ pounds per square inch}$$
$$< 24{,}000 \text{ pounds per square inch} \ldots \text{satisfactory}$$

REFERENCES

1. American Concrete Institute. *Building Code Requirements and Commentary for Reinforced Concrete* (ACI 318-05). Farmington Hills, MI, 2005.
2. American Society of Civil Engineers. *Minimum Design Loads for Buildings and Other Structures: ASCE 7-05.* New York, 2005.
3. Williams, A. *Structural Engineering License Review Problems and Solutions,* 6th ed. Dearborn, Chicago, 2007.
4. American Concrete Institute. *ACI Design Handbook.* Farmington Hills, MI, 1997.
5. Portland Cement Association. *PCA Column to ACI 318-02.* Skokie, IL, 2003.
6. Williams, A. *Design of Reinforced Concrete Structures,* 4th ed. Dearborn, Chicago, 2007.
7. Zia, P., et al. Estimating Prestress Losses. *Concrete International: Design and Construction.* Vol. 1, No. 6, June 1979, pp. 32–38.
8. Portland Cement Association. *Notes on ACI 318-05 Building Code Requirements for Reinforced Concrete.* Skokie, IL, 2005.
9. Lin, T. Y. Load Balancing Method for Design and Analysis of Prestressed Concrete Structure. *Proceedings of the American Concrete Institute.* Vol. 60, 1963, pp. 719–742.
10. Prestressed Concrete Institute. *PCI Design Handbook.* Chicago, 2004.
11. Concrete Society. *Post-Tensioned Flat Slab Design Handbook.* London, 1984.
12. American Institute of Steel Construction. *Steel Construction Manual,* 13th ed. Chicago, 2005.
13. American Institute of Steel Construction. *Basic Design Values Cards.* Chicago, 2005.
14. International Code Council. *International Building Code—2006.* Falls Church, VA, 2006.
15. American Forest and Paper Association. *National Design Specifications for Wood Construction With Commentary and Supplement (ANSI/AF&PA NDS-2005).* Washington, DC, 2005.
16. American Plywood Association. *Glued-Laminated Beam Design Tables.* Tacoma, WA, 2001.
17. Western Wood Products Association. *Western Lumber Span Tables.* Portland, OR, 2001.
18. American Plywood Association. *Diaphragms.* APA Research Report 138. Tacoma, WA, 2000.
19. Brandow, G. E. *UBC Diaphragm Requirements.* Structural Engineering Association of Southern California Design Seminar, Los Angeles, 1992.
20. Coil, J. *Subdiaphragms.* Structural Engineering Association of Southern California Design Seminar. Los Angeles, 1991.
21. American Concrete Institute. *Building Code Requirements for Masonry Structures (ACI 530-05).* Farmington Hills, MI, 2005.
22. Concrete Masonry Association of California and Nevada. *1997 Design of Reinforced Masonry Structures.* Citrus Heights, CA, 1997.

CHAPTER 3

Bridge Structures

PROBLEMS

3.1 The two-span bridge with a single central column support shown in Exhibit 3.1 is located in the vicinity of San Diego on a nonessential route. The soil profile at the site consists of a 35-foot layer of soft-to-medium clay.

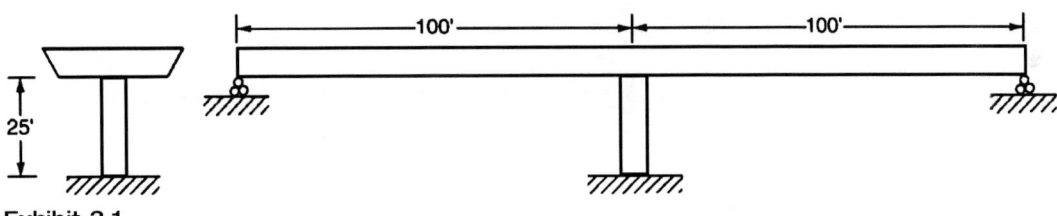

Exhibit 3.1

    a. The applicable acceleration coefficient to be used for seismic design is most nearly:
        a. 0.40
        b. 0.30
        c. 0.20
        d. 0.10

    b. What importance category should be assigned to the structure?
        a. Critical   c. Other
        b. Essential   d. Insufficient information

    c. What seismic performance zone should be assigned to the structure?
        a. 1
        b. 2
        c. 3
        d. 4

    d. The value of the site coefficient is most nearly:
        a. 1.1
        b. 1.2
        c. 1.3
        d. 1.5

e. What analysis procedure should be used?
   a. UL or SM
   b. UL
   c. SM
   d. MM

f. The column has a moment of inertia of 40 feet^4 and a modulus of elasticity of 432,000 kips per square foot and may be assumed fixed at the top and bottom. The stiffness of the column is most nearly:
   a. 12,970 kips per foot
   b. 13,070 kips per foot
   c. 13,170 kips per foot
   d. 13,270 kips per foot

g. The weight of the superstructure and tributary substructure has a constant value of 7 kips per foot. The fundamental period of the bridge in the longitudinal direction is most nearly:
   a. 0.34 second
   b. 0.34 second
   c. 0.36 second
   d. 0.37 second

h. The value of the elastic seismic response factor is most nearly:
   a. 0.8
   b. 1.0
   c. 1.2
   d. 1.3

i. The elastic seismic moment in the column due to the longitudinal seismic force is most nearly:
   a. 14,000 kip-feet
   b. 15,000 kip-feet
   c. 16,000 kip-feet
   d. 17,000 kip-feet

j. The reduced design moment in the column is most nearly:
   a. 4300 kip-feet
   b. 4500 kip-feet
   c. 4700 kip-feet
   d. 4900 kip-feet

3.2 Exhibit 3.2 shows the central bent in a regular two-span bridge located in the vicinity of Los Angeles on a strategic route. The 4-foot-diameter columns may be considered fixed at the top and bottom, and the axial force due to dead load at the bottom of each column is 800 kips. Each column is reinforced with 24 No. 14 deformed bars, Grade 60, and the concrete strength is 3.25 kips per square inch. The relevant column interaction diagram is shown in the figure.

a. Do slenderness effects have to be considered in designing the columns?
   a. Yes
   b. No

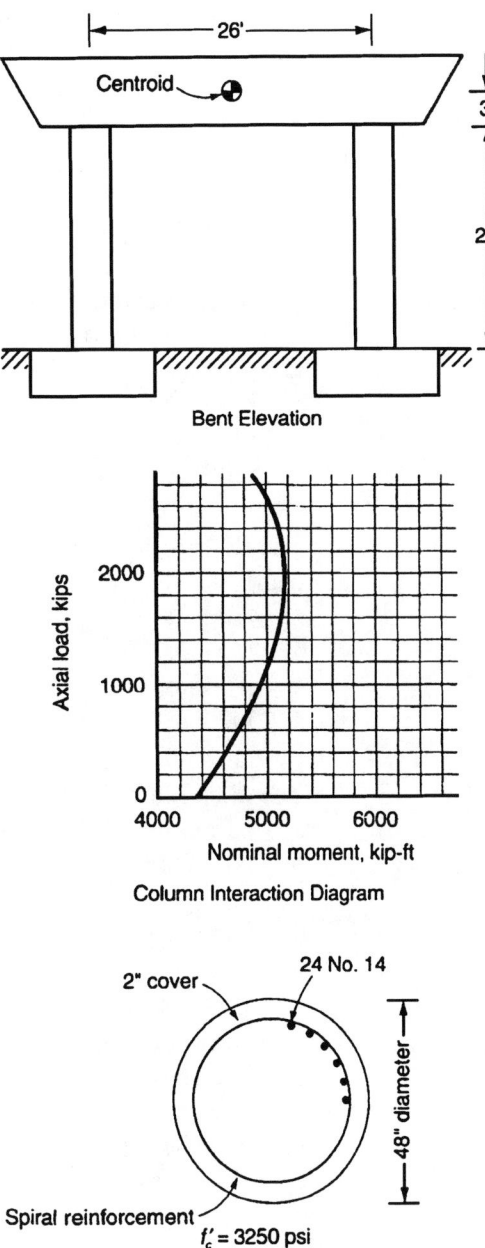

Exhibit 3.2

b. From the given axial force due to dead load and the interaction diagram provided, the maximum probable plastic hinging moment at the base of each column is most nearly:
   a. 6370 kip-feet
   b. 6470 kip-feet
   c. 6570 kip-feet
   d. 6670 kip-feet

c. If the overstrength plastic hinge capacities at the top and bottom of the column may be assumed equal, the maximum transverse shear force developed in a column in the two-column bent is most nearly:
   a. 630 kips
   b. 640 kips
   c. 650 kips
   d. 660 kips

d. The design shear strength provided by the concrete section outside of the end regions is most nearly:
   a. 150 kips
   b. 160 kips
   c. 170 kips
   d. 180 kips

e. The design shear strength required from shear reinforcement is most nearly:
   a. 470 kips
   b. 480 kips
   c. 490 kips
   d. 500 kips

f. The pitch required for a spiral of Number 6 reinforcement is most nearly:
   a. 4.5 inches
   b. 5.0 inches
   c. 5.5 inches
   d. 6.0 inches

g. The length of the end regions, over which special confinement reinforcement is required, is most nearly:
   a. 42 inches
   b. 44 inches
   c. 46 inches
   d. 48 inches

h. The minimum design shear strength provided by the concrete within the end regions of the column is most nearly:
   a. 0 kip
   b. 50 kips
   c. 90 kips
   d. 135 kips

i. The design shear strength required from shear reinforcement in the end regions of the left column is most nearly:
   a. 500 kips
   b. 520 kips
   c. 580 kips
   d. 620 kips

j. The pitch required for a spiral of No. 7 reinforcement in the end regions of the left column is most nearly:
   a. 3.00 inches
   b. 3.25 inches
   c. 3.50 inches
   d. 3.75 inches

3.3 Exhibit 3.3 shows part of the superstructure of a two-lane highway bridge with an effective span of 44 feet. The precast, pretensioned girders have a 28-day compressive strength of 6 kips per square inch, and the final prestressing force, acting at the position shown, has a magnitude of 300 kips, all losses having occurred before the flange is cast. The area of the stress-relieved strand is 2 square inches, and the strand has a specified tensile strength of 270 kips per square inch. Before placing the flange shuttering and casting the flange, each girder is propped at its center with a firm, rigid prop. The shuttering is supported by the girders and weighs 30 pounds per foot run per rib, and the 28-day compressive strength of the concrete in the flange is 3 kips per square inch. The design load produces a bending moment of $M_{max} = 260$ kip-feet in the composite section. The effects of differential shrinkage and additional superimposed dead load, added after the flange is cast, may be ignored.

Superstructure Details

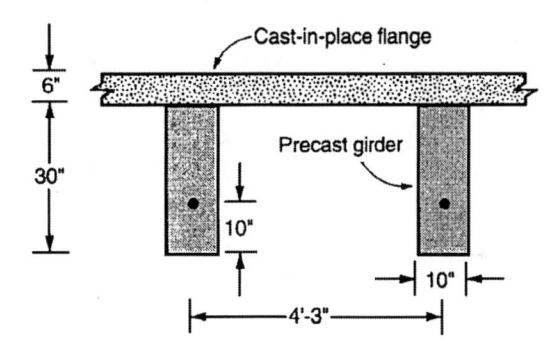

Exhibit 3.3

a. The stress in the bottom of the girder at the center of the span, attributable to the final prestressing force only, is most nearly:
   a. 1.80 kips per square inch
   b. 1.90 kips per square inch
   c. 2.00 kips per square inch
   d. 2.10 kips per square inch

b. The stress in the bottom of the girder at the center of the span, due to the self-weight of the girder only, is most nearly:
   a. −0.50 kip per square inch
   b. −0.60 kip per square inch
   c. −0.70 kip per square inch
   d. −0.80 kip per square inch

c. The stress in the bottom of the girder at the center of the span, due to the weight of the shuttering plus the weight of the flange concrete, is most nearly:
   a. 0.17 kip per square inch
   b. 0.19 kip per square inch
   c. 0.21 kip per square inch
   d. 0.23 kip per square inch

d. The section modulus of the composite section at the bottom of the girder is most nearly:
   a. 2500 inches^3
   b. 2600 inches^3
   c. 2700 inches^3
   d. 2800 inches^3

e. The stress produced in the bottom of the girder at the center of the span by the removal of the prop is most nearly:
   a. −0.25 kip per square inch
   b. −0.35 kip per square inch
   c. −0.45 kip per square inch
   d. −0.55 kip per square inch

f. The stress produced in the bottom of the girder at the center of the span by the removal of the shuttering is most nearly:
   a. 0.03 kip per square inch
   b. 0.04 kip per square inch
   c. 0.05 kip per square inch
   d. 0.06 kip per square inch

g. The maximum moment in the girder due to HS20 standard loading causes a stress at the bottom of the girder that is most nearly:
   a. 0.80 kip per square inch
   b. 0.90 kip per square inch
   c. 1.00 kip per square inch
   d. 1.10 kips per square inch

h. The final bottom fiber stress in the girder at the center of the span, due to all causes, is most nearly:
   a. 0.035 kip per square inch
   b. 0.045 kip per square inch
   c. 0.055 kip per square inch
   d. 0.065 kip per square inch

i. The design flexural capacity of the composite section is most nearly:
   a. 820 kip-feet
   b. 880 kip-feet
   c. 940 kip-feet
   d. 1000 kip-feet

3.4 Exhibit 3.4 shows the elevation of a two-span, posttensioned, box section, bridge superstructure. The properties of the box section, which are constant over the whole length of the bridge, are shown on the figure. The variation of the cable eccentricity is indicated on the figure, and the final prestressing force of 7000 kips may be considered constant over the whole length of the bridge.

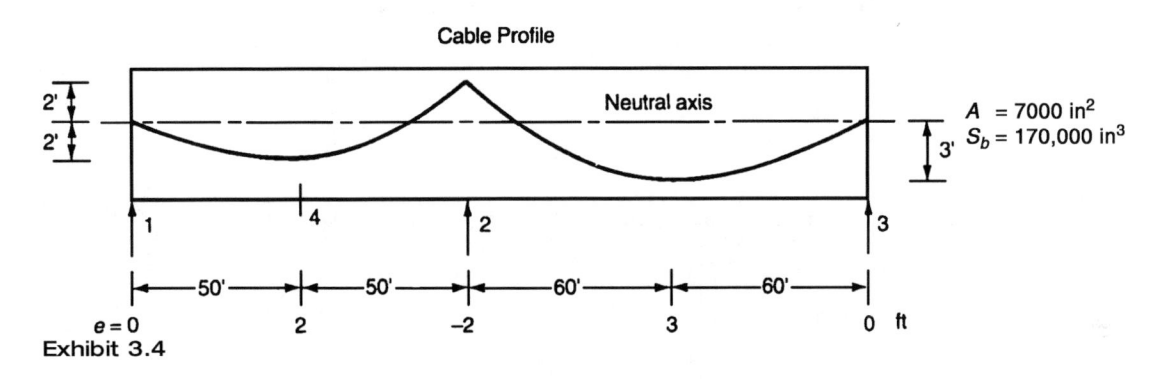

Cable Profile

$A = 7000$ in^2
$S_b = 170,000$ in^3

Exhibit 3.4

a. The primary bending moment at support 2, due to the final prestressing force, is most nearly:
   a. 10,000 kip-feet
   b. 14,000 kip-feet
   c. 18,000 kip-feet
   d. 21,000 kip-feet

b. The primary bending moment at section 4 is most nearly:
   a. 10,000 kip-feet
   b. 14,000 kip-feet
   c. 18,000 kip-feet
   d. 21,000 kip-feet

c. The moment at support 2, due to combined primary and secondary moments, is most nearly:
   a. 14,000 kip-feet
   b. 18,000 kip-feet
   c. 21,000 kip-feet
   d. 25,000 kip-feet

d. The secondary moment at support 2 is most nearly:
   a. 5400 kip-feet
   b. 7000 kip-feet
   c. 8600 kip-feet
   d. 10,800 kip-feet

e. The secondary reaction at support 1 is most nearly:
   a. 90 kips
   b. 110 kips
   c. 140 kips
   d. 170 kips

f. The secondary reaction at support 3 is most nearly:
   a. 90 kips
   b. 110 kips
   c. 140 kips
   d. 170 kips

g. The secondary reaction at support 2 is most nearly:
   a. 110 kips
   b. 140 kips
   c. 170 kips
   d. 200 kips

h. The secondary moment at section 4 is most nearly:
   a. 5400 kip-feet
   b. 7000 kip-feet
   c. 8600 kip-feet
   d. 10,800 kip-feet

i. The moment at section 4, due to combined primary and secondary moments, is most nearly:
   a. 5400 kip-feet
   b. 7000 kip-feet
   c. 8600 kip-feet
   d. 10,800 kip-feet

j. The stress in the bottom fiber of the section at support 2, due to combined primary and secondary moments, is most nearly:
   a. 0.30 kip per square inch
   b. 0.45 kip per square inch
   c. 0.60 kip per square inch
   d. 0.75 kip per square inch

3.5 Exhibit 3.5 shows the piling arrangement for a bridge abutment. The pile cap may be assumed rigid, and the soil under the cap carries no load. The piles are end-bearing and may be considered hinged at each end with negligible flexural stiffness. The value of *AE/L* is constant for all piles.

Determine the axial forces in the plies due to the loading indicated.

Exhibit 3.5

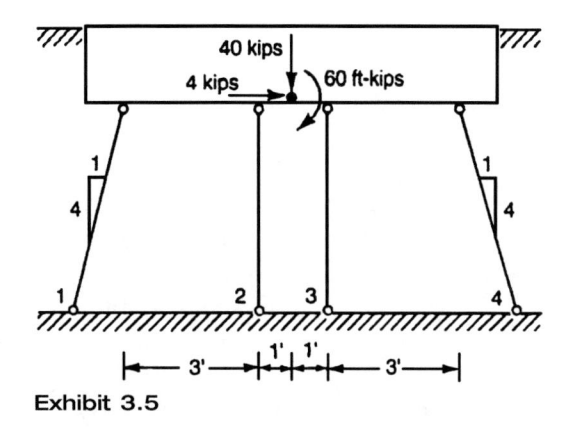

3.6 Exhibit 3.6 shows details of a simply supported steel composite bridge superstructure with full shear connection. The 9-inch concrete deck consists of 4.5 kips per square inch normal weight concrete. The W33 × 201 Grade 50 beams span 100 feet and have the following properties: A_s = 59.2 square inches, d = 33.7 inches, b_f = 15.7 inches.

Determine the design flexural strength of the composite section.

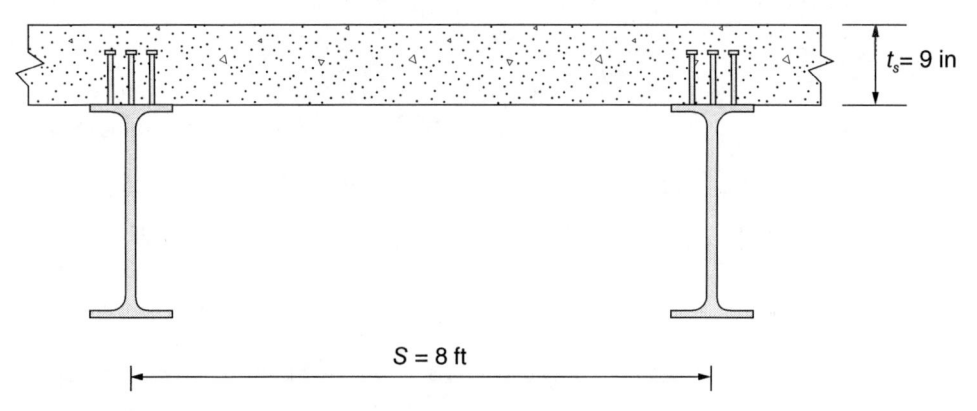

Exhibit 3.6

3.7 For the steel composite bridge superstructure shown in Exhibit 3.6, determine the total number of shear connectors required to provide full composite action at the strength I limit state. The stud connectors are ¾-inch diameter with a tensile strength of 65 kips per square inch.

3.8 The superstructure of a wood bridge consists of 10¾ × 30 inch glued-laminated 20F-V3 stringers supporting a glued-laminated panel deck. The deck panels adequately support the compression face of the stringers, and the moisture content of the stringers exceeds 16 percent. The stringers are simply supported over a span of 30 feet. Determine the design flexural strength of the stringers for live load.

SOLUTIONS

3.1 a. **a.** From AASHTO Section 3.10.2, the applicable acceleration coefficient for the San Diego area is $A = 0.4$.

b. **c.** From AASHTO Section 3.10.3, for a nonessential bridge, the importance category is IC = Other.

c. **d.** From Exhibit 3.1a, for a value of the acceleration coefficient exceeding 0.29, the relevant seismic performance zone is 4 (SPZ = 4).

Exhibit 3.1a Seismic performance zone

Acceleration Coefficient	Seismic Zone
$A \leq 0.09$	1
$0.09 < A \leq 0.19$	2
$0.19 < A \leq 0.29$	3
$0.29 < A$	4

d. **d.** From AASHTO Section 3.10.5, the relevant site coefficient for soil profile type III with a soft-to-medium clay layer with a depth of 35 feet is $S = 1.5$.

e. **a.** For a regular bridge in seismic zone 4 with an importance category of "Other," the required analysis procedure is UL or SM.

f. **d.** The stiffness of a column fixed at the top and bottom is given by
$$k_C = 12EI/H^3$$
$$= 12 \times 432{,}000 \times 40/25^3$$
$$= 13{,}271 \text{ kips per foot}$$

g. **c.** The total weight of the superstructure and tributary substructure is
$$W = wL$$
$$= 7 \times 200$$
$$= 1400 \text{ kips}$$

The longitudinal stiffness of the bridge is
$$k_C = 13{,}271/12$$
$$= 1106 \text{ kips per inch}$$

The fundamental period of the bridge in the longitudinal direction is given by
$$T_m = 0.32\sqrt{W/k_c}$$
$$= 0.32\sqrt{1400/1106}$$
$$= 0.36 \text{ second}$$

h. **a.** The value of the elastic seismic response coefficient is given by AASHTO Formula (C4.7.4.3.2c-3) as

$$C_{sm} = 1.2AS/T_m^{2/3}$$
$$= 1.2 \times 0.40 \times 1.5/(0.36)^{2/3}$$
$$= 1.42$$

For soil profile type III with $A \geq 0.3$, the maximum allowable value of C_{sm} is given by AASHTO Section 3.10.6.2 as

$$C_{sm} = 2.0A$$
$$= 2.0 \times 0.4$$
$$= 0.8 \ldots \text{governs}$$

i. **a.** The total elastic seismic shear is given by

$$V = WC_{sm}$$
$$= 1400 \times 0.8$$
$$= 1120 \text{ kips}$$

The elastic moment in the column is

$$M_E = VH/2$$
$$= 1120 \times 25/2$$
$$= 14{,}000 \text{ kip-feet}$$

j. **c.** The response modification factor for a single column is given in Exhibit 3.1b as $R = 3$. Hence, the reduced design moment in the column is

$$M_R = M_E/3$$
$$= 14{,}000/3$$
$$= 4667 \text{ kip-feet}$$

Exhibit 3.1b Response modification factors

		Importance Category	
Substructure	**Other**	**Essential**	**Critical**
Wall Type Pier:			
Strong Axis	2.0	1.5	1.5
Weak Axis	3.0	2.0	1.5
Reinforced Concrete Pile Bents:			
Vertical Piles Only	3.0	2.0	1.5
One or More Batter Piles	2.0	1.5	1.5
Single Columns:	3.0	2.0	1.5
Steel or Composite Pile Bents:			
Vertical Piles Only	5.0	3.5	1.5
One or More Batter Piles	3.0	2.0	1.5
Multiple Column Bent:	5.0	3.5	1.5

3.2 a. **b.** From AASHTO Section C5.7.4.3, the slenderness ratio of a circular column is given by kl_u/r
where
    k = effective length factor
    l_u = unsupported column length = 20 feet
    r = radius of gyration = $0.25 \times$ diameter = 1 foot

For an unbraced frame with both ends of the column fixed, the effective length factor is $k = 1.0$.

Hence, the slenderness ratio is

$$kl_u/r = 1.0 \times 20/1.0$$
$$= 20$$
$$< 22$$

Hence, the column is classified as a short column, and slenderness effects may be neglected.

b. a. From the interaction diagram, for an axial load of 800 kips, the nominal plastic hinging moment is

$$M_n = 4900 \text{ kip-feet}$$

In accordance with AASHTO Section 3.10.9.4.3b, the overstrength plastic hinge capacity is

$$M_{pr} = 1.3 M_n$$
$$= 1.3 \times 4900$$
$$= 6370 \text{ kip-feet}$$

c. d. The shear forces produced in the columns by the plastic hinges for a seismic force acting to the right are given by

$$V_{uL} = 2M_{prL}/H$$
$$= 2 \times 6370/20$$
$$= 637 \text{ kips}$$

$$V_{uR} = 2M_{prR}/H$$
$$= 2 \times 6370/20$$
$$= 637 \text{ kips}$$

The total shear force in the bent is

$$V_1 = V_{uL} + V_{uR}$$
$$= 637 + 637$$
$$= 1274 \text{ kips}$$

The axial forces produced in the columns by the plastic hinges are given by

$$P_{uL} = -[H_C(V_{uL} + V_{uR}) - (M_{prL} + M_{prR})]/B$$
$$= -[23(637 + 637) - (6370 + 6370)]/26$$
$$= -1127 + 490$$
$$= -637 \text{ kips}$$

$$P_{uR} = +[H_C(V_{uL} + V_{uR}) - (M_{prL} + M_{prR})]/B$$
$$= 1127 - 490$$
$$= 637 \text{ kips}$$

The axial forces produced in the columns by the dead load plus the plastic hinges are given by

$$P_L = P_D + P_{uL}$$
$$= 800 - 637$$
$$= 163 \text{ kips}$$

$$P_R = P_D + P_{uR}$$
$$= 800 + 637$$
$$= 1437 \text{ kips}$$

Using these revised axial forces, the nominal plastic hinging moments in the columns are obtained from the interaction diagram as

$$M_{nL} = 4450 \text{ kip-feet}$$
$$M_{nR} = 5100 \text{ kip-feet}$$

The corresponding overstrength plastic hinge capacities are given by

$$M_{prL} = 1.3 \times 4450$$
$$= 5785 \text{ kip-feet}$$

$$M_{prR} = 1.3 \times 5100$$
$$= 6630 \text{ kip-feet}$$

The shear forces produced in the columns by the revised plastic hinge capacities are given by

$$V_{uL} = 2M_{prL}/H$$
$$= 2 \times 5785/20$$
$$= 579 \text{ kips}$$

$$V_{uR} = 2M_{prR}/H$$
$$= 2 \times 6630/20$$
$$= 663 \text{ kips}$$

The total shear force in the bent is

$$V_2 = V_{uL} + V_{uR}$$
$$= 579 + 663$$
$$= 1242 \text{ kips}$$

The percentage change in total shear between Cycle 1 and Cycle 2 is

$$\Delta V = 100 (V_1 - V_2)/V_1$$
$$= 100 (1274 - 1242)/1274$$
$$= 2.5\%$$
$$< 10\%.$$

Hence, in accordance with AASHTO Section 3.10.9.4.3c, no further iterations are necessary, and the maximum probable shear force in the right column is

$$V_{max} = 663 \text{ kips}$$

d. **d.** In accordance with AASHTO Section 5.8.3.3 the design shear strength provided by the concrete, outside of the end regions, is

$$\phi V_c = 0.0632 \phi b_v d_v \sqrt{f'_c}$$

where

    ϕ = strength reduction factor = 0.90 from AASHTO Section 5.5.4.2.1
    b_v = column diameter = 48 inches
    d_v = distance from compressive fiber to centroid of reinforcement in opposite side of member
        = 24 + 0.637 × 20.4
        = 37 inches
    f'_c = concrete compressive strength = 3.25 kips per square inch

Hence,

$$\phi V_c = 0.632 \times 0.90 \times 48 \times 37 \times \sqrt{3.25}$$
$$= 182 \text{ kips}$$

e. **b.** The design shear strength required from the shear reinforcement is given by AASHTO Equation (5.8.3.3-1) as

$$\phi V_s = V_u - \phi V_c$$
$$= 663 - 182$$
$$= 481 \text{ kips}$$

f. **d.** To satisfy the requirements for lateral reinforcement in a compression member, in accordance with AASHTO Section 5.10.6.2, the center-to-center distance between spirals shall not exceed 6 inches, and the minimum volumetric ratio of the spiral reinforcement to the concrete core is given by AASHTO Equation (5.7.4.6-1) as

$$\rho_s = 0.45 \, (A_g/A_c - 1) f'_c/f_y$$
$$= A_v \pi (D_c - D_s)/sA_c$$

where

    A_g = gross area of column = 1810 square inches
    A_c = area of concrete core measured to outside of spiral = 1521 square inches
    D_c = diameter of core measured to outside of spiral = 44 inches
    D_s = diameter of spiral reinforcement = 0.75 inch
    A_v = area of spiral reinforcement = 0.44 square inch
    s = pitch of spiral reinforcement

Hence, the required pitch is given by

$$s = A_v f_y \, \pi(D_c - D_s)/0.45 A_c f'_c (A_g/A_c - 1)$$
$$= 0.44 \times 60{,}000 \, \pi(44 - 0.75)/[0.45 \times 1521 \times 3250 \, (1810/1521 - 1)]$$
$$= 8.5 \text{ inches}$$

$$s = 6 \text{ inches maximum} \ldots \text{governs}$$

To satisfy the requirements for shear strength in accordance with AASHTO Section 5.8.3.3, the required spiral pitch is given by AASHTO Equation (5.8.3.3-4) as

$$s = \phi A_v f_y d_v / \phi V_s$$
$$= 0.90 \times 2 \times 0.44 \times 60 \times 37/481$$
$$= 3.66 \text{ inches} \ldots \text{governs}$$
$$< 6.0 \text{ inches}$$

g. d. In accordance with AASHTO Section 5.10.11.4.1c, the length of the end regions is the larger of
   (i) 18 inches
   (ii) H/6 = 20 × 12/6 = 40 inches
   (iii) Column diameter = 48 inches ... governs

h. b. The final axial forces produced in the columns by the final values of the overstrength plastic hinge moments are given by

$$P_{uL} = -[H_C(V_{uL} + V_{uR}) - (M_{prL} + M_{prR})]/B$$
$$= -[23(579 + 663) - (5785 + 6630)]/26$$
$$= -1098 + 478$$
$$= -621 \text{ kips}$$

$$P_{uR} = [H_C(V_{uL} + V_{uR}) - (M_{prL} + M_{prR})]/B$$
$$= 1098 - 478$$
$$= 621 \text{ kips}$$

The final axial forces produced in the columns by the dead load plus the plastic hinges are given by

$$P_L = P_D + P_{uL}$$
$$= 800 - 621$$
$$= 179 \text{ kips}$$

$$P_R = P_D + P_{uR}$$
$$= 800 + 621$$
$$= 1421 \text{ kips}$$

The axial force value given by

$$A_c f'_c / 10 = 1521 \times 3.25/10$$
$$= 494 \text{ kips}$$
$$> P_L$$

Hence, in accordance with AASHTO Section 5.10.11.4.1c, the design shear strength of the concrete in the end region of the left column is given by

$$\phi V'_c = \phi V_c P_L / 494$$
$$= 182 \times 179/494$$
$$= 66 \text{ kips}$$

i. b. For the left column, the design shear strength required from the shear reinforcement is given by AASHTO Equation (5.8.3.3-1) as

$$\phi V_s = V_{uL} - \phi V'_c$$
$$= 579 - 66$$
$$= 513 \text{ kips}$$

j. **d.** The required pitch of confinement reinforcement is given by the smaller value obtained from AASHTO Equations (5.7.4.6-1) and (5.10.11.4.1d-1). Thus, for No. 7 reinforcement,

$$s = A_v f_y \pi(D_c - D_s)/0.45 A_c f'_c (A_g/A_c - 1)$$
$$= 0.60 \times 60{,}000\pi (44 - 0.875)/[0.45 \times 1521 \times 3250(1810/1521 - 1)]$$
$$= 11.6 \text{ inches}$$

or

$$s = A_v f'_c \pi(D_c - D_s)/0.12 A_c f'_c$$
$$= 0.60 \times 60{,}000\pi(44 - 0.875)/[0.12 \times 1521 \times 3250]$$
$$= 8.22 \text{ inches}$$

To satisfy the requirements for shear strength in accordance with AASHTO Section 5.8.3.3, the required spiral pitch is given by AASHTO Equation (5.8.3.3-4) as

$$s = \phi A_v f_y \, d/\phi V_s$$
$$= 0.90 \times 2 \times 0.60 \times 60 \times 37/513$$
$$= 4.67 \text{ inches}$$

Since the center-to-center distance between spirals shall not exceed 4 inches, in accordance with AASHTO Section 5.10.11.4.1e, the maximum pitch is

$$s = 4 \text{ inches}$$

3.3 a. **c.** The relevant properties of the precast girder are

$$A = 10 \times 30 = 300 \text{ inches}^2$$
$$S_b = 10 \times 30^2/6 = 1500 \text{ inches}^3$$
$$e = 30/2 - 10 = 5 \text{ inches} \ldots \text{lower kern position}$$

Hence, the stress in the bottom of the girder due to the prestressing force is

$$f_b = 2P_e/A$$
$$= 2 \times 300/300$$
$$= 2.0 \text{ kips per square inch}$$

b. **b.** The bending moment due to the girder self-weight is

$$M_G = 0.150 \times 300 \times 44^2 \times 12/(144 \times 8)$$
$$= 908 \text{ kip inches}$$

The stress in the bottom of the girder due to this moment is

$$f_b = -M_G/S_b$$
$$= -908/1500$$
$$= -0.605 \text{ kip per square inch}$$

c. **a.** The weight of the shuttering is

$$w_S = 0.030 \text{ kip per foot}$$

The weight of the flange concrete is

$$w_F = 0.15 \times A_F$$
$$= 0.15 \times 4.25 \times 0.5$$
$$= 0.319 \text{ kip per foot}$$

The total weight of the shuttering plus flange concrete is

$$w = w_S + w_F$$
$$= 0.030 + 0.319$$
$$= 0.349 \text{ kip per foot}$$

The central prop creates a two-span beam with spans of 22 feet as shown in Exhibit 3.3a(i), with a central reaction of

$$R_P = 10wl/8$$
$$= 10 \times 0.349 \times 22/8$$
$$= 9.60 \text{ kips}$$

The moment at the prop due to the shuttering plus flange concrete is

$$M_P = wl^2/8$$
$$= 1.5 \times 0.349 \times 22^2$$
$$= 253 \text{ kip-inches}$$

The stress in the bottom of the girder due to this moment is

$$f_b = M_P/S_b$$
$$= 253/1500$$
$$= 0.169 \text{ kip per square inch}$$

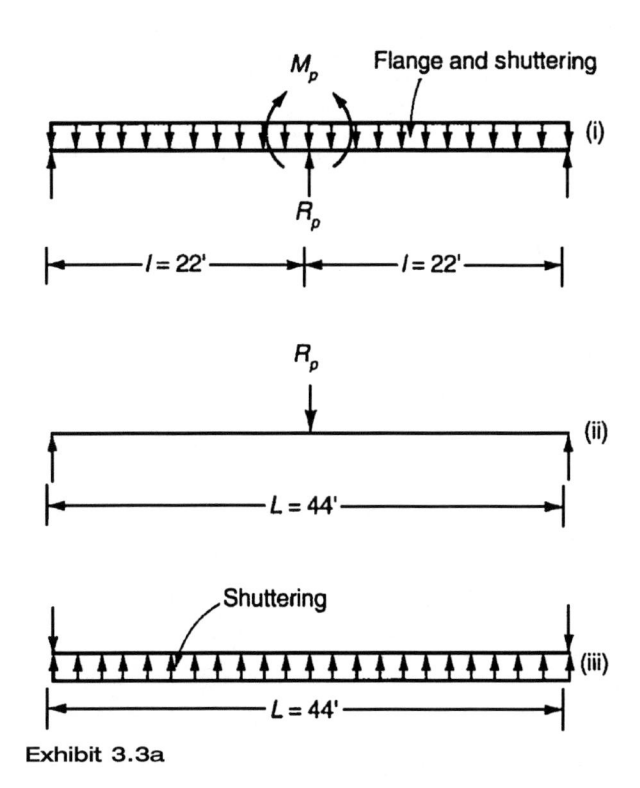

Exhibit 3.3a

d. **d.** The modular ratio for the cast-in-place flange and the precast girder is given by AASHTO Section C5.4.2.4 as

$$n = E_f/E_r$$
$$= \sqrt{f'_{c(\text{flange})}/f'_{c(\text{girder})}}$$
$$= \sqrt{3/6}$$
$$= 0.707$$

The effective compression flange width is given by AASHTO Sections 4.6.2.6.1 as the minimum of

(1) $b = L/4$
$= 44/4$
$= 11$ feet

(2) $b = b_w + 12h_f$
$= 10 + 12 \times 6$
$= 82$ inches

(3) $b = S$
$= 4.25 \times 12$
$= 51$ inches ... governs

The transformed flange width is

$$b_t = nb$$
$$= 0.707 \times 51$$
$$= 36 \text{ inches}$$

The section properties of the composite section are obtained as shown in Exhibit 3.3b. Hence,

$$\bar{y} = \Sigma Ay/\Sigma A$$
$$= 11{,}628/516$$
$$= 22.5 \text{ inches}$$

$$I_c = \Sigma I + \Sigma Ay^2 - \bar{y}^2 \Sigma A$$
$$= 63{,}836 \text{ inches}^4$$

Exhibit 3.3b Composite section properties

Part	A	y	I	Ay	Ay^2
Girder	300	15	22,500	4500	67,500
Flange	216	33	648	7128	235,224
Total	516		23,148	11,628	302,724

The section modulus at the bottom of the section is

$$S_{Cb} = I_C/\bar{y}$$
$$= 2833 \text{ inches}^3$$

e. **c.** Removal of the prop is equivalent to applying a downward load to the composite section at midspan equal in magnitude to the reaction in the prop, as shown in Exhibit 3.3a (ii). This produces the moment

$$M_R = R_P L/4$$
$$= 9.6 \times 44 \times 12/4$$
$$= 1267 \text{ kip-inches.}$$

The stress in the bottom of the girder due to this moment is

$$f_{Cb} = -M_r/S_{Cb}$$
$$= -1267/2833$$
$$= -0.447 \text{ kips per square inch}$$

f.   **a.** Removal of the shuttering is equivalent to applying an upward load to the composite section equal in magnitude to the weight of the shuttering, as shown in Exhibit 3.3a (iii). This produces the moment

$$M_S = w_S L^2/8$$
$$= 1.5 \times 0.03 \times 44^2$$
$$= 87.12 \text{ kip-inches}$$

The stress in the bottom of the girder due to this moment is

$$f_{Cb} = M_S/S_{Cb}$$
$$= 87.12/2833$$
$$= 0.031 \text{ kip per square inch}$$

The stress in the bottom of the girder due to the design moment is

$$f_{Cb} = -M_{max}/S_{Cb}$$
$$= -12 \times 260/2833$$
$$= -1.101 \text{ kips per square inch}$$

g.   **b.** The final bottom fiber stress in the girder is

$$f_b = 2.000 - 0.605 + 0.169 - 0.447 + 0.031 - 1.101$$
$$= 0.047 \text{ kip per square inch, compression}$$

h.   **d.** The effective prestress in the tendons after all losses is

$$f_{pe} = P_e/A_{ps}$$
$$= 300/2$$
$$= 150 \text{ kips per square inch}$$
$$> 0.5 f_{pu} \text{ ... AASHTO Section 5.7.3.1 is applicable}$$

The compression zone factor for 3 kips per square inch of concrete is

$$\beta_1 = 0.85 \text{ ... from AASHTO Section 5.7.2.2}$$

The prestressing steel factor is given by AASHTO Table 5.7.3.1.1-1 as

$$k = 0.38 \text{ ... for stress-relieved strand}$$

Assuming that the neutral axis lies within the flange for a section without non-prestressed tension reinforcement, the distance from the extreme compression fiber to the neutral axis is given by AASHTO Equation (5.7.3.1.1-4) as

$$c = A_{ps}f_{pu}/(0.85 f'_c \beta_1 b + k A_{ps} f_{pu}/d_p)$$
$$= 2 \times 270/(0.85 \times 3 \times 0.85 \times 51 + 0.38 \times 2 \times 270/26)$$
$$= 4.56 \text{ inches}$$
$$< h_f \text{ ... neutral axis is within the flange}$$

The depth of the equivalent rectangular stress block is given by AASHTO Section 5.7.2.2 as

$$a = \beta_1 c$$
$$= 0.85 \times 4.56$$
$$= 3.9 \text{ inches ... stress block is within the slab}$$

The stress in bonded tendons at ultimate load is given by AASHTO Equation (5.7.3.1.1-1) as

$$f_{ps} = f_{pu}(1 - kc/d_p)$$
$$= 270(1 - 0.38 \times 4.56/26)$$
$$= 252 \text{ kips per square inch}$$

The nominal flexural strength of the section is given by AASHTO Equation (5.7.3.2.2-1) as

$$M_n = A_{ps} f_{ps}(d_p - a/2)$$
$$= 2 \times 252(26 - 3.9/2)$$
$$= 12{,}120 \text{ kip-inches}$$

The strain in the prestressing tendons at the nominal flexural strength is

$$\epsilon_t = \epsilon_{cu}(d_p - c)/c$$
$$= 0.003(26 - 4.56)/4.56$$
$$= 0.014$$
$$> 0.005 \text{ ... section is tension controlled}$$

Hence, the resistance factor is given by AASHTO Section 5.5.4.2.1 as

$$\phi = 1.0$$

The design flexural capacity is

$$M_r = \Phi M_n$$
$$= 1.0 \times 12{,}120$$
$$= 12{,}120 \text{ kip-inches}$$
$$= 1010 \text{ kip-feet}$$

3.4 a. b. The primary moment at support 2, as shown in Exhibit 3.4a at (i), is

$$M_{2p} = Pe$$
$$= 7000 \times (-2)$$
$$= -14{,}000 \text{ kip-feet, sagging}$$

b. b. The primary moment at section 4, as shown in Exhibit 3.4a at (i), is

$$M_{4p} = Pe$$
$$= 7000 \times 2$$
$$= 14{,}000 \text{ kip-feet, hogging}$$

c. d. The drape of the parabolic cable is span 12 is

$$a = 2 + 2/2$$
$$= 3 \text{ feet}$$

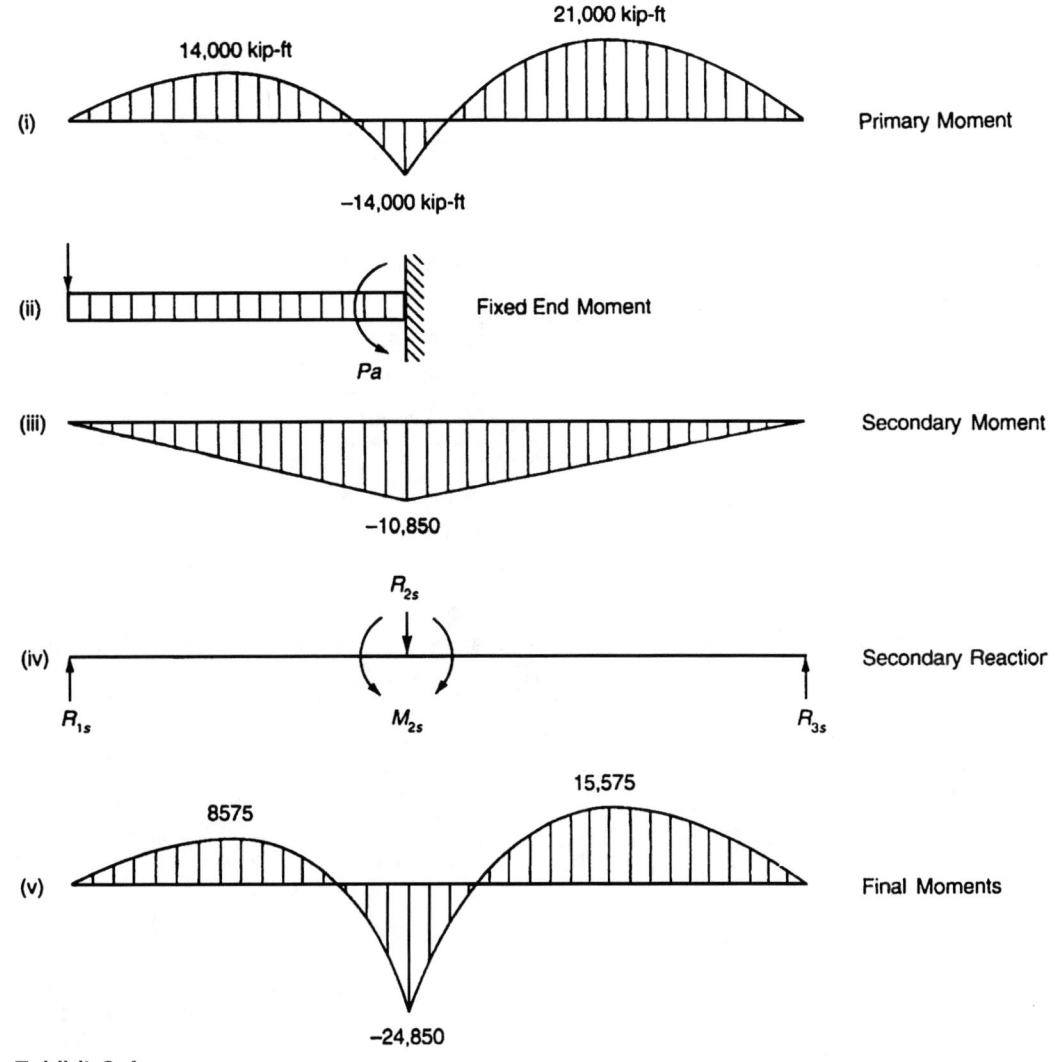

Exhibit 3.4a

(i) Primary Moment
14,000 kip-ft, 21,000 kip-ft, −14,000 kip-ft

(ii) Fixed End Moment — Pa

(iii) Secondary Moment — −10,850

(iv) Secondary Reaction — R_{1s}, R_{2s}, M_{2s}, R_{3s}

(v) Final Moments — 8575, 15,575, −24,850

The equivalent upward lateral balancing load is

$$w = 8Pa/L^2$$

Allowing for the hinge at support 1, the initial fixed-end moment at support 2, as shown in Exhibit 3.4a at (ii), is

$$\begin{aligned}M_{21}^F &= -wL^2/8, \text{ anticlockwise} \\ &= -Pa \\ &= -7000 \times 3 \\ &= -21{,}000 \text{ kip-feet}\end{aligned}$$

The drape of the parabolic cable in span 23 is

$$\begin{aligned}a &= 3 + 2/2 \\ &= 4 \text{ feet}\end{aligned}$$

Allowing for the hinge at support 3, the initial fixed-end moment at support 2 is

$$\begin{aligned}M_{23}^F &= Pa, \text{ clockwise} \\ &= 7000 \times 4 \\ &= 28{,}000 \text{ kip-feet}\end{aligned}$$

These initial fixed-end moments are distributed as shown in Exhibit 3.4b. Advantage is taken of the hinged supports to eliminate carryover to ends 1 and 3.

Exhibit 3.4b Moment distribution

Joint	1	2	2	3
Member	12	21	23	32
Relative EI/L		3/100	3/120	
Distribution factors		0.55	0.45	
Fixed-end moment		−21,000	28,000	
Distribution		−3,850	−3,150	
Final moments		−24,850	24,850	

The final moment at support 2, due to combined primary and secondary moment, is

$$M_{2e} = -24{,}850 \text{ kip-feet, sagging}$$

d. **d.** The secondary moment at support 2, as shown in Exhibit 3.4a at (iii), is

$$\begin{aligned} M_{2s} &= M_{2e} - M_{2p} \\ &= -24{,}850 + 14{,}000 \\ &= -10{,}850 \text{ kip-feet, sagging} \end{aligned}$$

e. **b.** The secondary reaction at support 1, as shown in Exhibit 3.4a at (iv), is given by

$$\begin{aligned} R_{1s} &= M_{2s}/L_{12} \\ &= -10{,}850/100 \\ &= -108.5 \text{ kips, upward} \end{aligned}$$

f. **a.** The secondary reaction at support 3, as shown in Exhibit 3.4a at (iv), is given by

$$\begin{aligned} R_{3s} &= M_{2s}/L_{23} \\ &= -10{,}850/120 \\ &= -90.4 \text{ kips, upward} \end{aligned}$$

g. **d.** The secondary reaction at support 2, as shown in Exhibit 3.4a at (iv), is given by

$$\begin{aligned} R_{2s} &= -(R_{1s} + R_{3s}) \\ &= 108.5 + 90.4 \\ &= 198.9 \text{ kips, downward} \end{aligned}$$

h. **a.** The secondary moment at section 4, as shown in Exhibit 3.4a at (iii), is

$$\begin{aligned} M_{4s} &= R_{1s}\, L_{12}/2 \\ &= M_{2s}/2 \\ &= -5425 \text{ kip-feet, sagging} \end{aligned}$$

i. c. The final moment at section 4, as shown in Exhibit 3.4a at (v), is given by

$$M_{4e} = M_{4p} + M_{4s}$$
$$= 14{,}000 - 5425$$
$$= 8575 \text{ kip-feet}$$

j. d. The final bottom fiber stress at section 2 is given by

$$f_{2b} = P/A + M_{2e}/S_b$$
$$= 7000/7000 - 24{,}850 \times 12/170{,}000$$
$$= -0.754 \text{ kip per square inch, tension}$$

3.5 The displacement of the cap and piles is a combination of horizontal translation, vertical translation, and rotation. The solution may be obtained by means of the displacement method of analysis. Adopting matrix notation, the vector of pile cap displacements is given by

$$[\Delta] = \begin{bmatrix} x \\ y \\ \theta \end{bmatrix}$$

The vector of pile axial forces is

$$[P] = \begin{bmatrix} P_1 \\ P_2 \\ P_3 \\ P_4 \end{bmatrix}$$

The vector of applied loads is

$$[W] = \begin{bmatrix} H \\ V \\ M \end{bmatrix} = \begin{bmatrix} 4 \\ 40 \\ 60 \end{bmatrix}$$

The diagonal matrix formed from the axial stiffnesses of the individual piles is

$$[\bar{S}] = AE/L \begin{bmatrix} 1 & 0 & 0 & 0 \\ 0 & 1 & 0 & 0 \\ 0 & 0 & 1 & 0 \\ 0 & 0 & 0 & 1 \end{bmatrix}$$

The elements of the transformation matrix are the axial deformations produced in the piles by unit value of each pile displacement imposed in turn, and the transformation matrix is given by

$$[T] = \begin{bmatrix} -\tfrac{1}{4} & 1 & -4 \\ 0 & 1 & -1 \\ 0 & 1 & 1 \\ \tfrac{1}{4} & 1 & 4 \end{bmatrix}$$

The complete stiffness matrix for the whole system is

$$[S] = [T]^T [\overline{S}] [T] = AE/L \begin{bmatrix} \frac{1}{8} & 0 & 2 \\ 0 & 4 & 0 \\ 2 & 0 & 34 \end{bmatrix}$$

The vector of pile cap displacements is given by

$$[\Delta] = [S]^{-1} [W] = L/AE \begin{bmatrix} 64 \\ 10 \\ -2 \end{bmatrix}$$

The vector of pile axial forces is given by

$$[P] = [\overline{S}] [T] [\Delta] = \begin{bmatrix} 2 \\ 12 \\ 8 \\ 18 \end{bmatrix}$$

3.6 The effective width of the concrete deck is given by AASHTO Section 4.6.2.6.1 as the least of

$$b = L/4$$
$$= 100 \times 12/4$$
$$= 300 \text{ inches}$$

$$\text{or } b = 12t_s + b_f/2$$
$$= 12 \times 9$$
$$= 108 \text{ inches}$$

$$\text{or } b = S$$
$$= 8 \times 12$$
$$= 96 \text{ inches } \ldots \text{ governs}$$

The depth of the stress block in a fully composite beam is given by

$$D_p = F_y A_s / 0.85 f'_c b$$
$$= 50 \times 59.2/(0.85 \times 4.5 \times 96)$$
$$= 8.06 \text{ inches } \ldots \text{ stress block is within the slab}$$

The moment arm between centroids of the tensile force and the compressive force is

$$\ell_a = d/2 + t_s - D_p/2$$
$$= 33.7/2 + 9 - 8.06/2$$
$$= 21.82 \text{ inches}$$

The plastic moment of resistance is

$$M_p = F_s A_s \ell_a$$
$$= 50 \times 59.2 \times 21.82/12$$
$$= 5382 \text{ kip-feet}$$

The ratio of depth of the stress block to overall depth of the composite section is

$$D_p/D_t = 8.06/42.7$$
$$= 0.19$$
$$> 0.1$$

Hence, AASHTO Equation (6.10.7.1.2-2) governs, and the nominal flexural resistance of the section is

$$M_n = M_p(1.07 - 0.7D_p/D_t)$$
$$= 5382(1.07 - 0.7 \times 0.19)$$
$$= 5043 \text{ kip-feet}$$

The top flange of the steel beam is anchored to the deck slab by shear connectors and is considered continuously braced. Hence, the resistance factor for flexure is given by AASHTO Section 6.5.4.2 as

$$\phi_f = 1.0$$

The design flexural resistance of the composite section is given by AASHTO Equation (6.10.7.1.1-1) as

$$M_u = \phi_f M_n$$
$$= 5043 \text{ kip-feet}$$

3.7 The total horizontal shear to be resisted by the shear connectors at the strength limit state is obtained from AASHTO Section 6.10.10.4.2 as the lesser of

$$P = 0.85 f'_c b t_s$$
$$= 0.85 \times 4.5 \times 96 \times 9$$
$$= 3305 \text{ kips}$$
$$\text{or } P = A_s F_y$$
$$= 59.2 \times 50$$
$$= 2960 \text{ kips ... governs}$$
$$< 3305 \text{ kips}$$

The modulus of elasticity of the concrete deck is given by AASHTO Equation (C5.4.2.4-1) as

$$E_c = 1820(f'_c)^{0.5}$$
$$= 1820(4.5)^{0.5}$$
$$= 3861 \text{ kips per square inch}$$

The nominal shear strength of a ¾-inch diameter stud shear connector is given by AASHTO Section 6.10.10.4.3 as the lesser of

$$Q_n = 0.5 A_{sc}(f'_c E_c)^{0.5}$$
$$= 0.5 \times 0.44(4.5 \times 3861)^{0.5}$$
$$= 29 \text{ kips}$$
$$\text{or } Q_n = A_{sc} F_u$$

$$= 0.44 \times 65$$
$$= 28.60 \text{ kips ... governs}$$
$$< 29 \text{ kips}$$

For full composite action, the total number of studs required on the beam is given by AASHTO Section 6.10.10.4.1 as

$$2n = 2P/\phi_{sc}Q_n$$
$$= 2 \times 2960/(0.85 \times 28.6)$$
$$= 244 \text{ studs}$$

3.8 The base design bending stress of the 20F-V3 stringer is obtained from AASHTO Table 8.4.1.2.3-1 as

$$F_{bo} = 5.08 \text{ kips per square inch}$$

The nominal design value in bending is
$$F_b = F_{bo}C_M C_F C_D C_T$$

where
- C_T = time effect factor given in AASHTO Section 8.4.4.5
  = 0.8
- C_M = wet service factor given in AASHTO Table 8.4.4.3-2
  = 0.8
- C_D = deck factor
  = 1.0
- C_F = size factor given by AASHTO Equation (8.4.4.2-3)
  $= (1291.5/bdL)^{1/a}$
  $= [1291.5/(10.75 \times 30 \times 30)]^{1/0.1}$
  = 0.82

Hence, the adjusted design value in bending is
$$F_b = 5.08 \times 0.8 \times 0.82 \times 0.8$$
$$= 2.67 \text{ kips per square inch}$$

The section modulus of the stringer is
$$S = bd^2/6$$
$$= 10.75 \times 30^2/6$$
$$= 1613 \text{ inches}^3$$

- C_s = stability factor
  = 1.0 ... compression face continuously supported
- ϕ = resistance factor
  = 0.85 ... for flexure

The nominal flexural resistance of the stringer is given by AASHTO Equation (8.6.2-1) as

$$M_n = F_b S C_S$$
$$= 2.67 \times 1613 \times 1.0/12$$
$$= 359 \text{ kip-feet}$$

The design flexural resistance is given by AASHTO Equation (8.6.1-1) as

$$M_r = \phi M_n$$
$$= 0.85 \times 359$$
$$= 305 \text{ kip-feet}$$

REFERENCES

1. American Association of State Highway and Transportation Officials. *Bridge Design Specifications,* 3rd ed. 2004 with 2005 and 2006 Interim Revisions. Washington, DC, 2004.
2. Williams, A. *The Analysis of Indeterminate Structures.* Macmillan, London, 1967.
3. American Institute of Steel Construction. *Moments, Shears, and Reactions: Continuous Highway Bridge Tables.* Chicago, 1959.
4. Jenkins, W. M. Influence Line Computations for Structures With Members of Varying Flexural Rigidity Using the Electronic Digital Computer. *Structural Engineer.* Vol. 39, September 1961.
5. Wang, C. K. Matrix Analysis of Statically Indeterminate Trusses. *Proceedings of the American Society of Civil Engineers.* Vol. 85 (ST4), April 1959.
6. Portland Cement Association. *Influence Lines Drawn as Deflection Curves.* Skokie, IL, 1948.
7. Thadani, B. N. Distribution of Deformation Method for the Construction of Influence Lines. *Civil Engineering and Public Works Review.* Vol. 51, June 1956.
8. Lee, S. L., and Patel, P. C. The Bar-Chain Method of Analyzing Truss Deformations. *Proceedings of the American Society of Civil Engineers.* Vol. 86 (ST3), May 1960.
9. Williams, A. The Determination of Influence Lines for Bridge Decks Monolithic With Their Piers. *Structural Engineer.* Vol. 42, May 1964.
10. Morice, P. B., and Little, G. *The Analysis of Right Bridge Decks Subjected to Abnormal Loading.* Cement and Concrete Association, London, 1956.
11. West, R. *Recommendations on the Use of Grillage Analysis for Slab and Pseudo-Slab Bridge Decks.* Cement and Concrete Association, London, 1973.
12. Loo, Y. C., and Cusens, A. R. A Refined Finite Strip Method for the Analysis of Orthotropic Plates. *Proceedings of the Institution of Civil Engineers.* Vol. 48, January 1971.
13. Davis, J. D., Somerville, I. J., and Zienkiewicz, O. C. Analysis of Various Types of Bridges by the Finite Element Method. *Proceedings of the Conference on Developments in Bridge Design and Construction, Cardiff, March 1971.* Crosby Lockwood, London, 1972.
14. Westergaard, H. M. Computation of Stresses in Bridge Slabs Due to Wheel Loads. *Public Roads.* March 1930.
15. Building Science Safety Council. *NEHRP Recommended Provisions for Seismic Regulations for New Buildings and Other Structures: Part 2, Commentary.* Washington, DC, 2000.

16. Paz, M. *Structural Dynamics.* Van Nostrand Reinhold, New York, 1991.
17. Federal Highway Administration. *Seismic Design and Retrofit Manual for Highway Bridges.* Washington, DC, 1987.
18. Hewlett-Packard Company. *HP-48G Calculator Reference Manual.* Corvallis, OR, 1994.
19. American Concrete Institute. *Building Code Requirements and Commentary for Reinforced Concrete (ACI 318-05).* Detroit, MI, 2005.
20. Freyermuth, C. L. *Design of Continuous Highway Bridges With Precast, Prestressed Concrete Girders.* Portland Cement Association, Skokie, IL, 1969.
21. Freyermuth, C. L., and Shoolbred, R. A. *Post-Tensioned, Prestressed Concrete.* Portland Cement Association, Skokie, IL, 1967.
22. The Concrete Society. *Post-Tensioned Flat-Slab Design Handbook.* London, 1984.
23. Nash, G. F. J. *Steel Bridge Design Guide: Composite Universal Beam Simply Supported Span.* Constructional Steel Research and Development Organization, Croydon, UK, 1984.

CHAPTER 4

Foundations and Retaining Structures

PROBLEMS

4.1 The stub column shown in Exhibit 4.1 supports a steel column and base plate with the indicated loads. The stub column is 18 inches square and is reinforced with Grade 60 deformed bars. Concrete compressive strength is 2000 pounds per square inch. The stub column is subjected to axial load only and is effectively braced against side sway by the floor slab.

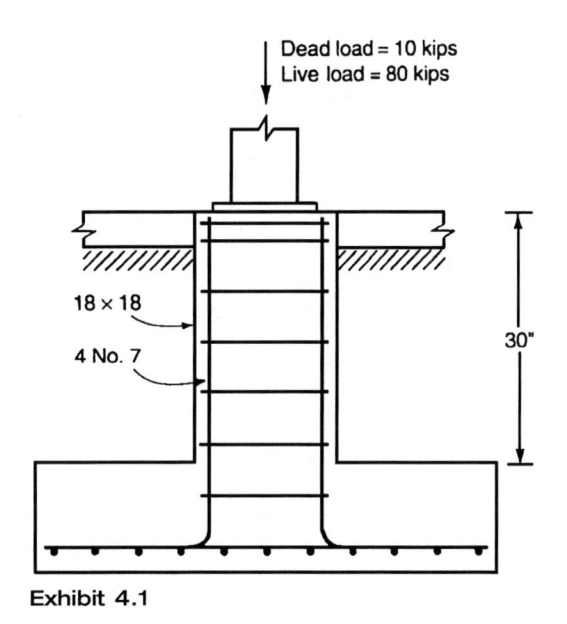

Exhibit 4.1

a. What is the effective length factor of the stub column?
  a. 0.65
  b. 0.80
  c. 1.00
  d. 1.20

b. The slenderness ratio is most nearly:
   a. 6
   b. 7
   c. 8
   d. 9

c. The design axial load strength of the stub column is most nearly:
   a. 300 kips
   b. 320 kips
   c. 340 kips
   d. 360 kips

d. The allowable reduced effective area of the section is most nearly:
   a. 160 square inches
   b. 170 square inches
   c. 180 square inches
   d. 190 square inches

e. The minimum allowable reinforcement area in the stub column is most nearly:
   a. 1.6 square inches
   b. 1.7 square inches
   c. 1.8 square inches
   d. 1.9 square inches

f. The minimum allowable size of lateral ties is:
   a. No. 3
   b. No. 4
   c. No. 5
   d. No. 6

g. The maximum allowable tie spacing is most nearly:
   a. 10 inches
   b. 12 inches
   c. 14 inches
   d. 16 inches

4.2 The fill behind the retaining wall in Exhibit 4.2 has a unit weight of 110 pounds per cubic foot with an equivalent fluid pressure of 30 pounds per square foot per foot. Passive earth pressure may be assumed equivalent to a fluid pressure of 300 pounds per square foot per foot, and the coefficient of friction at the underside of the base is 0.4. All concrete has a compressive strength of 3000 pounds per square inch, and reinforcement consists of Grade 60 deformed bars. For Questions **a** to **i**, neglect the effects of the shear key.

a. The value of the overturning moment about the toe is most nearly:
   a. 52,000 pound-feet
   b. 54,000 pound-feet
   c. 56,000 pound-feet
   d. 58,000 pound-feet

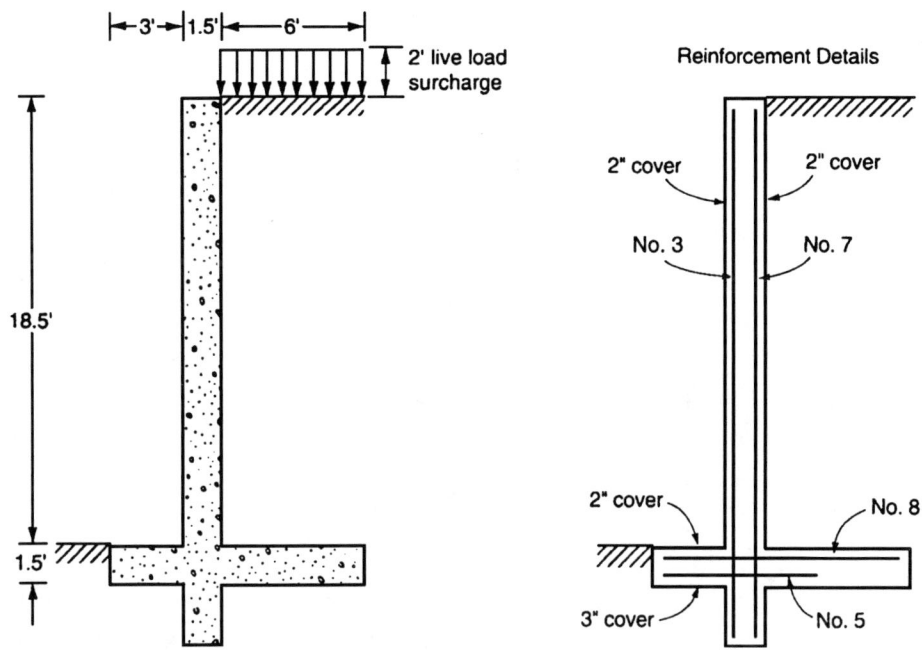

Exhibit 4.2

b. The distance of the resultant vertical load from the toe is most nearly:
   a. 5.5 feet
   b. 6.0 feet
   c. 6.5 feet
   d. 7.0 feet

c. The factor of safety against overturning is most nearly:
   a. 2.3
   b. 2.5
   c. 2.7
   d. 2.9

d. The earth pressure at the toe is most nearly:
   a. 3220 pounds per square foot
   b. 3270 pounds per square foot
   c. 3320 pounds per square foot
   d. 3370 pounds per square foot

e. The earth pressure at the heel is most nearly:
   a. 0 pounds per square foot
   b. 250 pounds per square foot
   c. 350 pounds per square foot
   d. 450 pounds per square foot

f. The maximum spacing of No. 8 bars in the heel is most nearly:
   a. 11 inches
   b. 12 inches
   c. 13 inches
   d. 14 inches

g. The maximum spacing of No. 5 bars in the toe is most nearly:
   a. 11 inches
   b. 12 inches
   c. 13 inches
   d. 14 inches

h. The maximum spacing of No. 7 bars in the stem is most nearly:
   a. 5.5 inches
   b. 6.0 inches
   c. 6.5 inches
   d. 7.0 inches

i. The maximum spacing of No. 3 bars in the front face of the stem is most nearly:
   a. 13 inches
   b. 12 inches
   c. 11 inches
   d. 10 inches

4.3 The ties for the anchored bulkhead shown in Exhibit 4.3 are located 4 feet from the top of the sheetpiling and are spaced at 15-foot centers. The end of the tie is secured to two anchor piles raked as indicated. The active earth pressure may be assumed equivalent to a fluid pressure of 30 pounds per square foot per foot, and passive pressure may be assumed equivalent to a fluid pressure of 400 pounds per foot per foot.

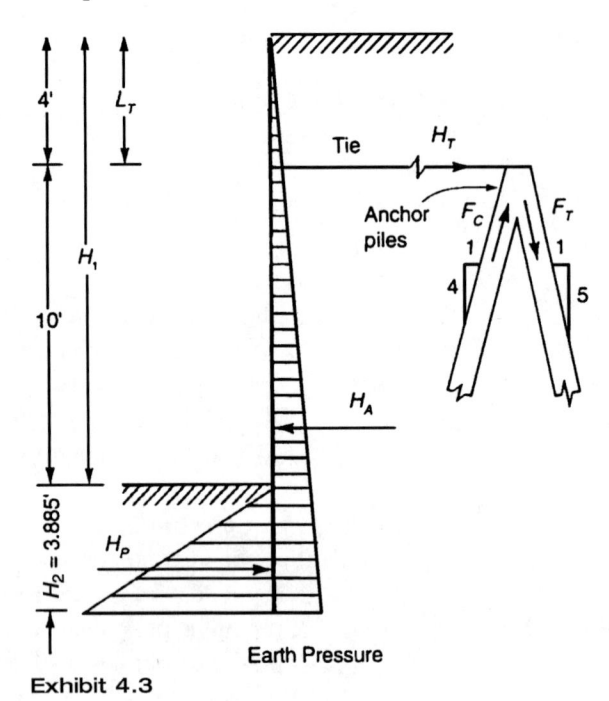

Exhibit 4.3

a. The value of the total active pressure on the back of the wall is most nearly:
   a. 2400 pounds per foot
   b. 3000 pounds per foot
   c. 3900 pounds per foot
   d. 4800 pounds per foot

b. The value of the total passive pressure on the front of the wall is most nearly:
   a. 1800 pounds per foot
   b. 2400 pounds per foot
   c. 3000 pounds per foot
   d. 3900 pounds per foot

c. The moment of all forces about the tie point is most nearly:
   a. 0 pounds feet per foot
   b. 200 pounds feet per foot
   c. 400 pounds feet per foot
   d. 600 pounds feet per foot

d. The force in the tie is most nearly:
   a. 1800 pounds per foot
   b. 2400 pounds per foot
   c. 3000 pounds per foot
   d. 3900 pounds per foot

e. The distance from the top of the wall to the position of maximum shear is most nearly:
   a. 0 feet
   b. 4 feet
   c. 9 feet
   d. 11 feet

f. The maximum shear in the wall is most nearly:
   a. 1240 pounds
   b. 1340 pounds
   c. 1440 pounds
   d. 1540 pounds

g. The distance from the top of the wall to the position of maximum moment is most nearly:
   a. 0 feet
   b. 4 feet
   c. 9 feet
   d. 11 feet

h. The maximum moment in the wall is most nearly:
   a. 5600 pounds feet per foot
   b. 5700 pounds feet per foot
   c. 5800 pounds feet per foot
   d. 5900 pounds feet per foot

i. The force in the compression anchor pile is most nearly:
   a. 60.5 kips
   b. 61.1 kips
   c. 61.7 kips
   d. 62.4 kips

j. The force in the tension anchor pile is most nearly:
   a. 60.5 kips
   b. 61.1 kips
   c. 61.7 kips
   d. 62.4 kips

4.4 The fill behind the retaining wall in Exhibit 4.4 has a unit weight of 110 pounds per cubic foot with an equivalent fluid pressure of 30 pounds per square foot per foot. The live load surcharge behind the wall is equivalent to an additional two feet of fill. The 12-inch square piles are spaced at 5 feet on center longitudinally.

a. The value of the total factored vertical load is most nearly:
   a. 120 kips
   b. 130 kips
   c. 140 kips
   d. 150 kips

b. The value of the total factored horizontal load is most nearly:
   a. 47 kips
   b. 57 kips
   c. 67 kips
   d. 77 kips

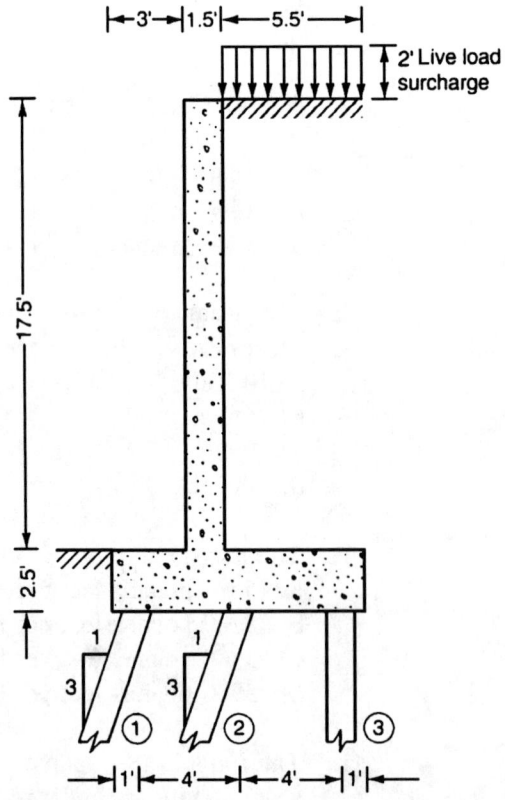

Exhibit 4.4

c. The height of the elastic center above the base is most nearly:
   a. 12 feet
   b. 15 feet
   c. 18 feet
   d. 21 feet

d. The horizontal distance of the elastic center from the toe is most nearly:
   a. 6 feet
   b. 7 feet
   c. 8 feet
   d. 9 feet

e. The moment of the factored loads about the elastic center is most nearly:
   a. 248 kip-feet
   b. 258 kip-feet
   c. 268 kip-feet
   d. 278 kip-feet

f. The force in pile 1 due to translation of the pile cap is most nearly:
   a. 75 kips
   b. 83 kips
   c. 91 kips
   d. 99 kips

g. The force in pile 1 due to rotation of the pile cap is most nearly:
   a. 68 kips
   b. 73 kips
   c. 78 kips
   d. 83 kips

h. The total force in pile 2 is most nearly:
   a. 69 kips
   b. 96 kips
   c. 135 kips
   d. 165 kips

i. The total force in pile 3 is most nearly:
   a. 53 kips
   b. 58 kips
   c. 63 kips
   d. 68 kips

4.5 The tilt-up concrete wall panel shown in Exhibit 4.5 is located 6 inches from the property line and supports an axial load of 5 kips per linear foot. The wall panel is supported on the eccentric footing indicated, and the ground floor slab is tied to the wall panel with Grade 60 reinforcement to limit the soil-bearing pressure to 3000 pounds per square foot.

a. Determine the required depth, H, of the footing below the level of the ground floor slab to produce a maximum soil bearing pressure of 3000 pounds per square foot.

b. Determine the area of the reinforcement required to tie the wall panel to the ground floor slab.

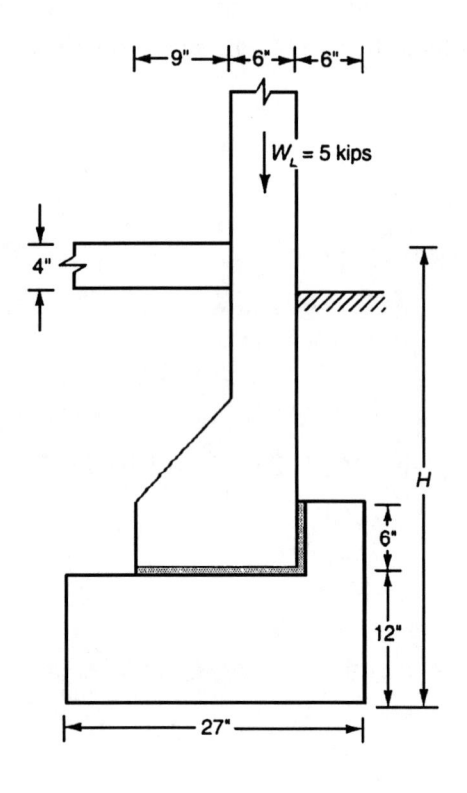

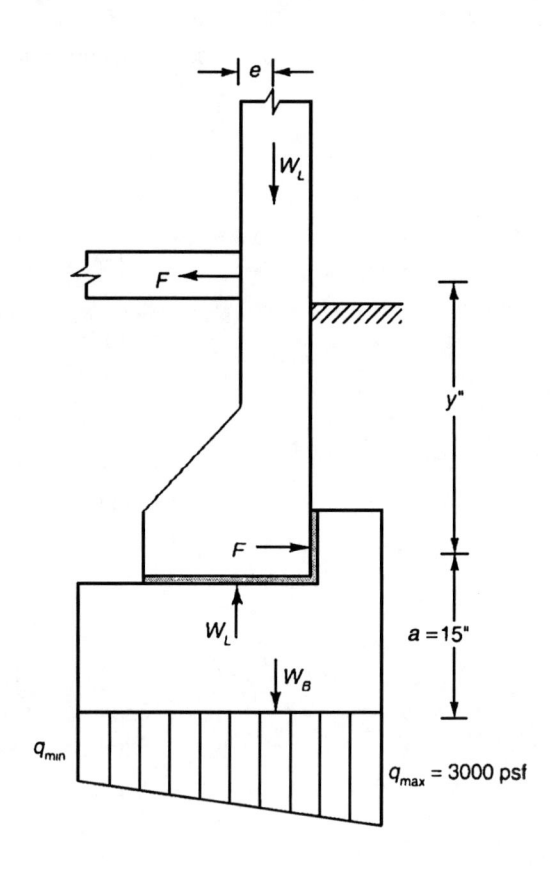

64 Chapter 4 Foundations and Retaining Structures

Exhibit 4.5

SOLUTIONS

4.1 **a. c.** From ACI Section 10.12.1, the effective length factor for a column braced against side sway is

$$k = 1.0$$

b. a. The radius of gyration, in accordance with ACI Section 10.11.2, is

$$r = 0.3c$$
$$= 0.3 \times 18$$
$$= 5.4 \text{ inches}$$

The slenderness ratio, in accordance with ACI Section 10.11.5, is

$$kl_u/r = 1.0 \times 30/5.4$$
$$= 5.6$$

c. d. For a column with lateral ties, the design axial load strength is given by ACI Equation (10-2) as

$$\phi P_n = 0.80\phi[0.85f'_c(A_g - A_{st}) + f_y A_{st}]$$
$$= 0.8 \times 0.65[0.85 \times 2 (324 - 2.4) + 60 \times 2.4]$$
$$= 359 \text{ kips}$$

d. **a.** The ratio of design load strength to applied ultimate load is

$$\phi P_n/P_u = 359/(1.2 \times 10 + 1.6 \times 80)$$
$$= 2.6$$
$$> 2$$

From ACI Section 10.8.4, the reduced effective area of the section is

$$A'_g = A_g/2$$
$$= 324/2$$
$$= 162 \text{ square inches}$$

e. **a.** The minimum allowable reinforcement area, in accordance with ACI Section 10.9.1, is

$$\rho_{min} = 0.01 A'_g$$
$$= 0.01 \times 162$$
$$= 1.62 \text{ square inches}$$

f. **a.** The minimum allowable size of lateral ties for enclosing bars smaller than No. 10 is given by ACI Section 7.10.5.1 as Number 3.

g. **c.** ACI Section 7.10.5.2 specifies a tie spacing not greater than

$$16 d_b = 16 \times 7/8 = 14 \text{ inches}$$
$$48 d_t = 48 \times 3/8 = 18 \text{ inches}$$
$$c = 18 \text{ inches}$$

4.2 a. **a.** The horizontal service loads acting on the wall are given by

$$H_A = \text{lateral pressure from backfill}$$
$$= 30 \times 20^2/2$$
$$= 6000 \text{ pounds}$$
$$H_L = \text{lateral pressure from surcharge}$$
$$= 2 \times 30 \times 20$$
$$= 1200 \text{ pounds}$$

Overturning moment about the toe is

$$M_O = H_A \times H/3 + H_L \times H/2$$
$$= 6000 \times 20/3 + 1200 \times 20/2$$
$$= 52{,}000 \text{ pound-feet}$$

b. **c.** The vertical service loads acting on the wall are given by

$$W_w = \text{weight of stem wall}$$
$$= 150 \times 1.5 \times 18.5$$
$$= 4163 \text{ pounds}$$

66   Chapter 4   Foundations and Retaining Structures

W_B = weight of base
    = 150 × 1.5 × 10.5
    = 2363 pounds

W_S = weight of backfill
    = 110 × 6 × 18.5
    = 12,210 pounds

W_L = weight of surcharge
    = 110 × 6 × 2
    = 1320 pounds

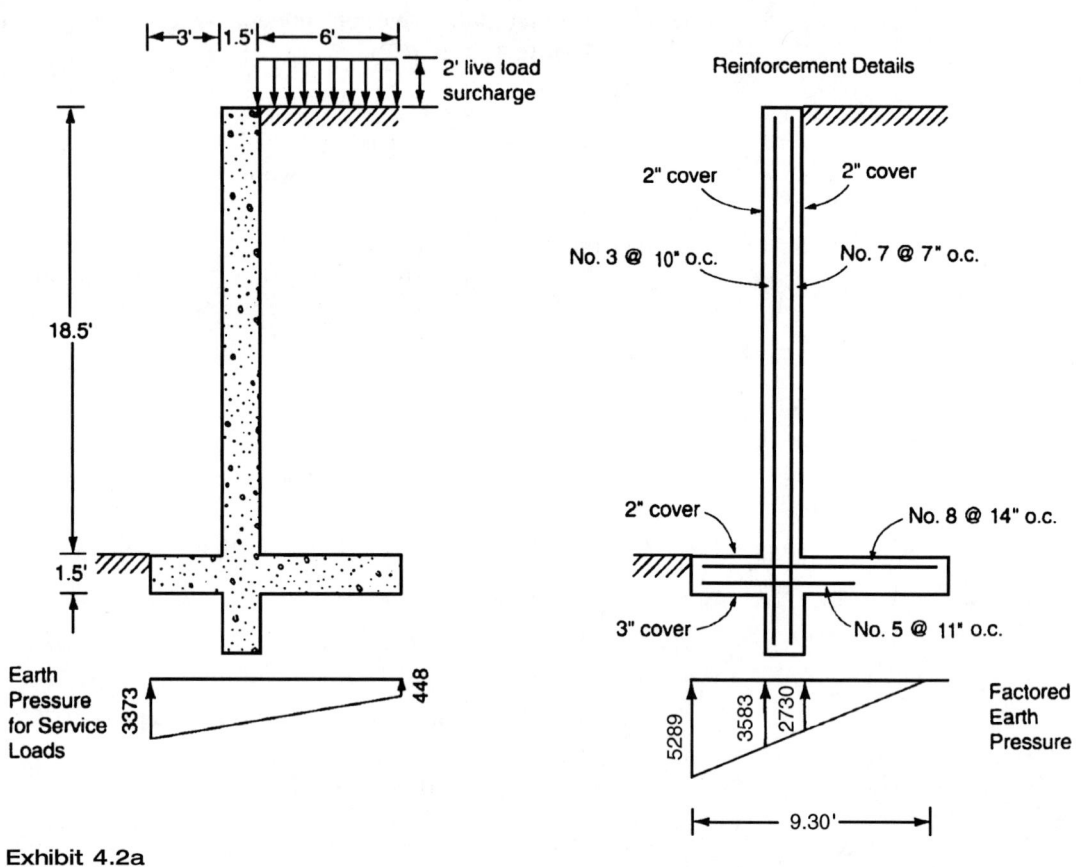

Exhibit 4.2a

The total vertical load is

$$\Sigma W = 20{,}056$$

The distance of the resultant vertical load from the toe is

$$\bar{x} = (4163 \times 3.75 + 2363 \times 5.25 + 12{,}210 \times 7.5 + 1320 \times 7.5)/20{,}056$$
$$= 6.5 \text{ feet}$$

   c. **b.**   The factor of safety against overturning is

$$\bar{x}\Sigma W / M_O = 6.5 \times 20{,}056 / 52{,}000$$
$$= 2.5$$

d. d. The eccentricity of the applied loads about the toe is given by

$$e' = (\bar{x}\Sigma W - M_O)/\Sigma W$$
$$= (6.5 \times 20{,}056 - 52{,}000)/20{,}056$$
$$= 3.91 \text{ feet}$$

The eccentricity about the base centroid is

$$e = 10.5/2 - 3.91$$
$$= 1.34 \text{ feet}$$
$$< L_B/6 \ldots \text{no tension under the base}$$

The pressure under the toe is

$$q_{\text{toe}} = \Sigma W(1 + 6e/L_B)/BL_B$$
$$= 20{,}056 \,(1 + 6 \times 1.34/10.5)/(1 \times 10.5)$$
$$= 3373 \text{ pounds per square foot}$$

e. d. The pressure under the heel is

$$q_{\text{heel}} = \Sigma W \,(1 - 6e/L_B)/BL_B$$
$$= 448 \text{ pounds per square foot}$$

f. d. The factored total overturning moment about the toe is

$$\gamma M_O = 1.6 \times 52{,}000$$
$$= 83{,}200 \text{ pound-feet}$$

The factored total vertical load is

$$\Sigma \gamma W = 1.2(W_W + W_B + W_S) + 1.6 W_L$$
$$= 4996 + 2836 + 14{,}652 + 2112$$
$$= 24{,}596 \text{ pounds}$$

The factored restoring moment is

$$\gamma M_R = 4996 \times 3.75 + 2836 \times 5.25 + 14{,}652 \times 7.5 + 2112 \times 7.5$$
$$= 159{,}354 \text{ pound-feet}$$

The eccentricity of the factored loads about the toe is given by

$$e' = (\gamma M_R - \gamma M_O)/\Sigma \gamma W$$
$$= 3.10 \text{ feet}$$
$$< L_B/3 \ldots \text{outside middle third}$$

The pressure under the toe is given by

$$q_{\text{toe}} = 2\Sigma \gamma W/3e'$$
$$= 2 \times 24{,}596/(3 \times 3.10)$$
$$= 5289 \text{ pounds per square foot}$$

The pressure distribution under the base is shown in Exhibit 4.2a. The maximum factored bending moment in the heel is

$$M_u = 3(\gamma W_S + \gamma W_L + 6\gamma W_B/10.5) - 2730 \times 4.80^2/6$$
$$= 3(14{,}652 + 2112 + 1621) - 10{,}483$$
$$= 44{,}672 \text{ pound-feet}$$

The required reinforcement ratio is derived from ACI Section 10.2, and is

$$\rho = 0.85\, f'_c\left(1-\sqrt{1-2K/0.765 f'_c}\right)/f_y$$
$$= 0.36 \text{ percent}$$

The maximum reinforcement ratio for a tension-controlled section is given by

$$\rho_t = 1.36 \text{ percent}$$
$$> \rho \ldots \text{satisfactory}$$

The minimum reinforcement ratio is given by ACI Section 7.12 as

$$\rho_{min} = 0.18 \text{ percent of the gross area}$$
$$< \rho \ldots \text{satisfactory}$$

Thus, the required reinforcement area is

$$A_S = \rho b d$$
$$= 0.0036 \times 12 \times 15.5$$
$$= 0.67 \text{ square inches per foot}$$

Shear is not critical, and No. 8 bars at a spacing of 14 inches provide a reinforcement area of

$$A'_s = 0.68 \text{ square inch per foot}$$
$$> A_s \ldots \text{satisfactory}$$

g. **b.** The maximum factored bending moment in the toe is

$$M_u = 3^2(3583 + 2 \times 5289)/6 - 1.5 \times 3\gamma W_B/10.5$$
$$= 21{,}242 - 1215$$
$$= 20{,}026 \text{ pound-feet}$$

The required reinforcement ratio is

$$\rho = 0.177 \text{ percent} \ldots \text{satisfactory}$$

The required reinforcement area is

$$A_S = 0.00177 \times 12 \times 14.69$$
$$= 0.31 \text{ square inch per foot}$$

Shear is not critical, and No. 5 bars at a spacing of 12 inches provide a reinforcement area of

$$A'_s = 0.31 \text{ square inch per foot}$$
$$= A_s \ldots \text{satisfactory}$$

h. **d.** The maximum factored bending moment in the stem is

$$M_u = 1.6(30 \times 18.5^3)/6 + 60 \times 18.5^2/2)$$
$$= 67{,}081 \text{ pound-feet}$$

The required reinforcement ratio is

$$\rho = 0.55 \text{ percent} \ldots \text{satisfactory}$$

The required reinforcement area is

$$A_s = 0.0055 \times 12 \times 15.5$$
$$= 1.03 \text{ square inches per foot}$$

Shear is not critical, and No. 7 bars at a spacing of 7 inches provide a reinforcement area of

$$A'_s = 1.03 \text{ square inches per foot}$$
$$= A_s \ldots \text{satisfactory}$$

i. d. Based on the gross concrete area, the reinforcement ratio required for the vertical reinforcement in the outer face of the wall is given by ACI Section 14.3 as

$$\rho = 0.0012/2 \text{ for bars not larger than No. 5}$$

The required vertical reinforcement area is

$$A_s = 0.0012 \times 12 \times 18/2$$
$$= 0.13 \text{ square inch per foot}$$

Providing No. 3 bars at a spacing of 10 inches gives

$$A'_s = 0.13 \text{ square inch per foot}$$
$$= A_s \ldots \text{satisfactory}$$

4.3 a. d. The total active pressure on the back of the wall is

$$H_A = p_A (H_1 + H_2)^2/2$$
$$= 30(14 + 3.885)^2/2$$
$$= 4798 \text{ pounds per foot}$$

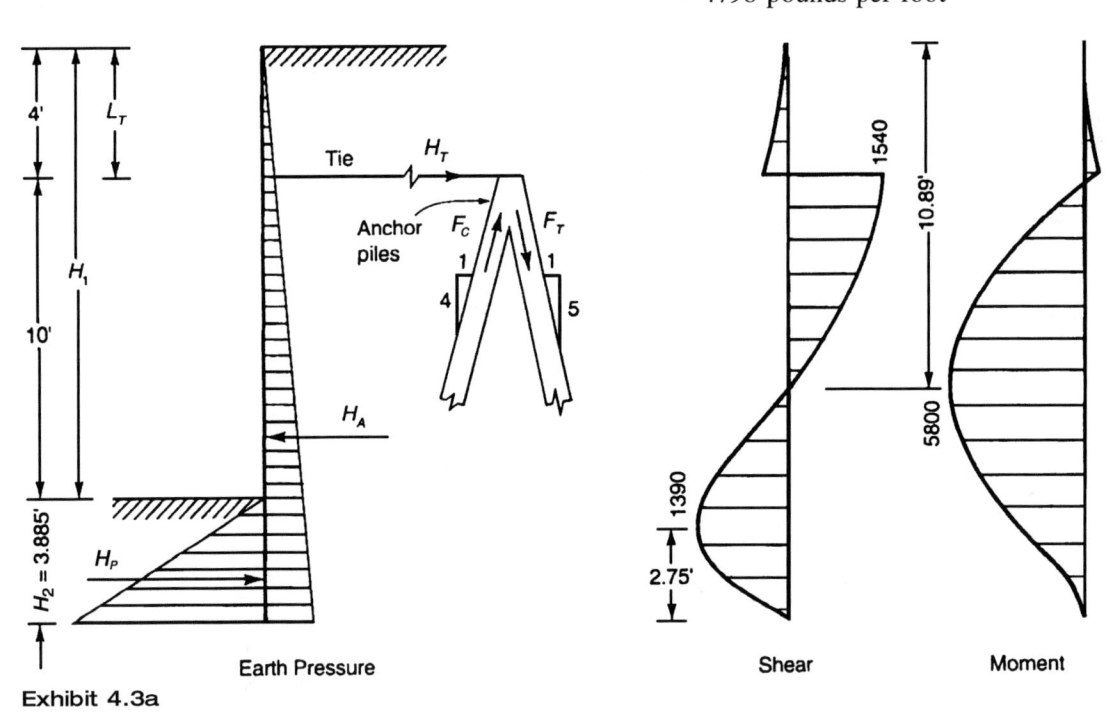

Exhibit 4.3a

b. c. The total passive pressure on the front of the wall is

$$H_P = p_p H_2^2/2$$
$$= 400 \times 3.885^2/2$$
$$= 3018 \text{ pounds per foot}$$

c. **a.** The bending moment at the tie point is

$$M_T = H_A[2(H_1 + H_2)/3 - L_T] - H_P(H_1 - L_T + 2H_2/3)$$
$$= 4798(2 \times 17.885)/3 - 4) - 3018(10 + 2 \times 3.885/3)$$
$$= 7 \text{ pound-feet per foot}$$

d. **a.** The force in the tie is

$$H_T = H_A - H_P$$
$$= 4798 - 3018$$
$$= 1780 \text{ pounds}$$

e. **b.** The maximum shear occurs at the tie point which is 4 feet from the top of the sheetpiling.

f. **d.** The maximum shear is given by

$$V_{max} = H_T - p_A L_T^2/2$$
$$= 1780 - 240$$
$$= 1540 \text{ pounds per foot}$$

g. **d.** The shear force at a distance x from the top of the sheetpiling when x exceeds L_T is

$$V = H_T - p_A x^2/2$$
$$= 1780 - 15x^2$$

The maximum moment occurs when $V = 0$. Hence,

$$x = \sqrt{1780/15}$$
$$= 10.89 \text{ feet}$$

h. **c.** The maximum moment is

$$M_{max} = H_T(x - L_T) - p_A x^3/6$$
$$= 1780(10.89 - 4) - 30 \times 10.89^3/6$$
$$= 5800 \text{ pound-feet per foot}$$

i. **b.** The force in the compression pile is

$$F_C = 15H_T/(\sin \theta_1 + \cos \theta_1 \tan \theta_2)$$
$$= 15 \times 1780/(\sin 14° + \cos 14° \tan 11.3°)$$
$$= 15 \times 1780/(0.243 + 0.970 \times 0.200)$$
$$= 61{,}094 \text{ pounds}$$

j. **a.** The force in the tension pile is

$$F_T = 15H_T/(\sin \theta_2 + \cos \theta_2 \tan \theta_1)$$
$$= 15 \times 1780/(\sin 11.3° + \cos 11.3° \tan 14°)$$
$$= 15 \times 1780/(0.196 + 0.981 \times 0.250)$$
$$= 60{,}524 \text{ pounds}$$

4.4 a. **a.** The factored vertical loads for a 5-foot length of wall are given by

$$W_W = \text{factored weight of stem wall}$$
$$= 5 \times 1.2 \times 150 \times 1.5 \times 17.5/1000$$
$$= 23.63 \text{ kips}$$

$$W_B = \text{factored weight of base}$$
$$= 5 \times 1.2 \times 150 \times 2.5 \times 10/1000$$
$$= 22.50 \text{ kips}$$

$$W_S = \text{factored weight of backfill}$$
$$= 5 \times 1.2 \times 110 \times 5.5 \times 17.5/1000$$
$$= 63.53 \text{ kips}$$

$$W_L = \text{factored weight of surcharge}$$
$$= 5 \times 1.6 \times 110 \times 5.5 \times 2/1000$$
$$= 9.68 \text{ kips}$$

$$W = \text{total factored vertical load}$$
$$= 23.63 + 22.50 + 63.53 + 9.68$$
$$= 119.33 \text{ kips}$$

b. **b.** The factored horizontal loads for a 5-foot length of wall are given by:

$$H_A = \text{factored lateral pressure from backfill}$$
$$= 5 \times 1.6 \times 30 \times 20^2/2000$$
$$= 48.00 \text{ kips}$$

$$H_L = \text{factored lateral pressure from surcharge}$$
$$= 5 \times 1.6 \times 2 \times 30 \times 20/1000$$
$$= 9.60 \text{ kips}$$

$$F = \text{total factored horizontal load}$$
$$= 48.00 + 9.60$$
$$= 57.60 \text{ kips}$$

c. **c.** The height of the elastic center above the base is

$$L_y = 3 \times 6$$
$$= 18 \text{ feet}$$

d. **d.** The horizontal distance of the elastic center from the toe is

$$L_x = 9 \text{ feet}$$

e. **d.** The clockwise moment of the factored loads about the elastic center is

$$M = 11.33H_A + 8H_L - 5.25W_W - 4W_B - 1.75(W_S + W_L)$$
$$= 544 + 77 - 124 - 90 - 128$$
$$= 278 \text{ kip-feet}$$

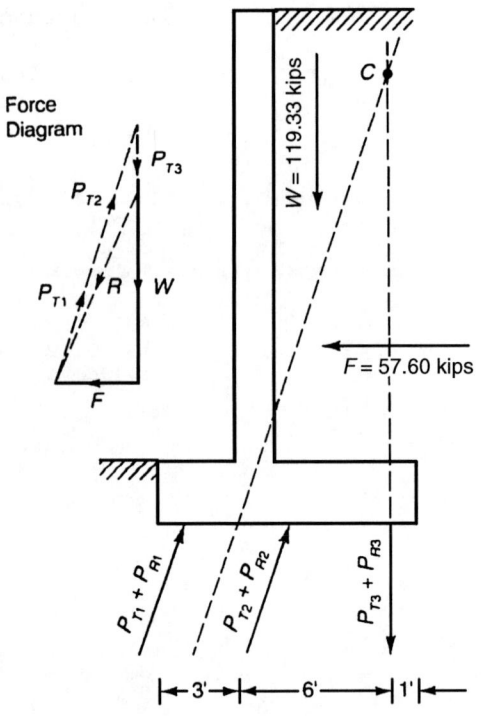

Exhibit 4.4a

f. **c.** Resolving forces, the axial force in pile 1 due to translation of the pile cap is given by

$$P_{T1} = F\sqrt{1+3^2}/2$$
$$= 57.60\sqrt{10}/2$$
$$= 91.07 \text{ kips} \ldots \text{compression}$$

g. **b.** The distance from the elastic center perpendicular to each pile is

$$r_1 = r_2 = 2 \cos 18.43°$$
$$= 1.90 \text{ feet}$$
$$r_3 = 0$$
$$\Sigma r^2 = 2(1.90)^2$$
$$= 7.2$$

The axial force in pile 1 due to rotation of the pile cap is

$$P_{R1} = Mr_1/\Sigma r^2$$
$$= 278 \times 1.90/7.2$$
$$= 73.36 \text{ kips} \ldots \text{tension}$$

h. **d.** The axial force in pile 2 due to translation is

$$P_{T2} = 91.07 \text{ kips} \ldots \text{compression}$$

The axial force in pile 2 due to rotation is

$$P_{R2} = 73.36 \text{ kips} \ldots \text{compression}$$

The total axial force in pile 2 is

$$P_2 = P_{T2} + P_{R2}$$
$$= 91.07 + 73.36$$
$$= 164.43 \text{ kips} \ldots \text{compression}$$

i. a. The axial force in pile 3 due to translation is

$$P_{T3} = W - 3F$$
$$= 119.33 - 3 \times 57.60$$
$$= -53.47 \text{ kips} \ldots \text{tension}$$

The axial force in pile 3 due to rotation is

$$P_{R3} = Mr_3/\Sigma r^2$$
$$= 0$$

The total axial force in pile 3 is

$$P_3 = P_{T3} + P_{R3}$$
$$= -53.47 \text{ kips} \ldots \text{tension}$$

4.5 **a.** The forces acting on the footing are shown in Exhibit 4.5 and assume a uniform distribution of pressure under the wall and between the side of the wall and the footing. The axial load in the wall panel is

$$W_L = 5 \text{ kips per foot}$$

The weight of the base is

$$W_B = 0.15(2.25 \times 1 + 0.5 \times 0.5)$$
$$= 0.375 \text{ kips per foot}$$

The total vertical load is

$$R = W_L + W_B$$
$$= 5.375 \text{ kips per foot}$$

The eccentricity between the axial load in the wall panel and the center of pressure under the wall is

$$e = 15/2 - 6/2$$
$$= 4.5 \text{ inches}$$

The clockwise couple produced by the axial load in the wall is

$$M_W = eW_L$$

This is balanced by the counterclockwise couple

$$M_F = yF$$

Equating the two couples gives

$$F = eW_L/y$$
$$= 4.5 \times 5/y$$
$$= 22.5/y \text{ kips}$$

The maximum pressure under the base is given by

$$q_{max} = R/A + Fa/S$$
$$= 5.375/2.25 + 22.5 \times 1.25/0.844y$$
$$= 3.0 \text{ kips per square foot}$$

Hence,

$$y = 54.5 \text{ inches}$$

The required depth of the footing below the level of the ground floor slab is

$$H = y + 15 + 4/2$$
$$= 71.5 \text{ inches}$$

b. The area of tensile reinforcement required in the floor slab is

$$A_S = 1.6F/\phi f_y$$
$$= 1.6 \times 22.5/(0.9 \times 60 \times 54.5)$$
$$= 0.012 \text{ square inch per foot}$$

REFERENCES

1. American Concrete Institute. *Building Code Requirements and Commentary for Structural Concrete (ACI 318-05)*. Farmington Hills, MI, 2005.
2. American Institute of Steel Construction. *Steel Construction Manual,* 13th ed. Chicago, 2005.
3. Williams, A. *Design of Reinforced Concrete Structures,* 4th ed. Kaplan® AEC Education, Chicago, 2008.
4. American Concrete Institute. *Buiding Code Requirements for Masonry Structures (ACI 530-05)*. Farmington Hills, MI, 2005.
5. Construction Industry Research and Information Association. *A Comparison of Quay Wall Design Methods* (CIRIA Technical Note 54). London, 1974.
6. Terzaghi, K. Anchored bulkheads. *Transactions American Society of Civil Engineers*. Vol. 119, New York, 1954.
7. Westergaards, H. M. The resistance of pile groups. *Engineering Construction*, New York, May 1918.
8. Vetter, C. P. Design of pile foundations. *Transactions American Society of Civil Engineers*. Vol. 64, New York, 1938.
9. Allen, A. H. *Reinforced Concrete Design to CP 100*. Cement and Concrete Association, London, 1974.
10. British Standards Institutions. *BS 8110: Structural Use of Concrete*. London, 1985.

CHAPTER 5

Hydraulics

PROBLEMS

5.1 A water district that serves a suburban community needs a new water transmission main from an existing reservoir. The line will be 5.1 miles long. The reservoir elevation is +280 feet, and the elevation at the final discharge point—a nonpressurized storage tank in town—is +100 feet. No pumping will be provided. It is proposed to build a single line of new, smooth, clean, cast-iron pipe. The Darcy friction factor is assumed to be 0.021. The present demand is 1 mgd.

a. The smallest appropriate commercially available pipe size for the line is
  a. 6 in.
  b. 8 in.
  c. 10 in.
  d. 12 in.

b. Is the initially assumed value for f appropriate?
  a. The initial value for f is too small.
  b. The initial value for f is appropriate.
  c. The initial value for f is too large.
  d. The appropriateness of f cannot be determined from the available data.

c. The Manning n that could apply to this new line is nearest to
  a. 0.010
  b. 0.012
  c. 0.014
  d. 0.016

5.2 The pipeline installation shown in Exhibit 5.2 is used to fill trucks with water. The 25 cm line is 30 m long; the 15 cm line, A, is 3 m long; the 15 cm line, B, is 15 m long. The Darcy-Wiesbach friction factor f is 0.02. Neglect minor losses.

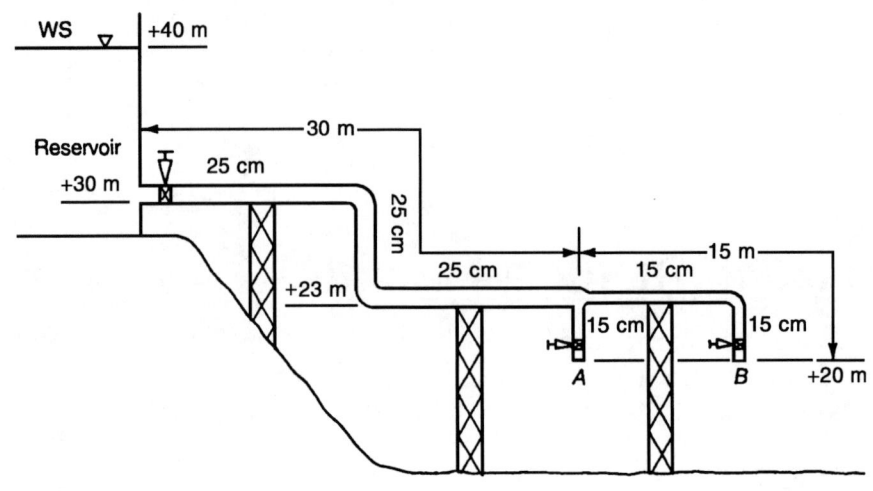

Exhibit 5.2

When all gate valves are fully open, the total discharge that can be delivered by the system is most nearly

a. 0.35 m³/sec
b. 0.40 m³/sec
c. 0.45 m³/sec
d. 0.50 m³/sec

5.3 An existing dam was included in a recent field study, and the following hydraulic and structural features were noted:

- The dam crest is at 1000 ft elevation.

- The roller gate is located on the crest of the spillway to control surcharge storage.

- The turbine conduit outlet is at 815 ft elevation.

Exhibit 5.3 presents a diagram of this system.

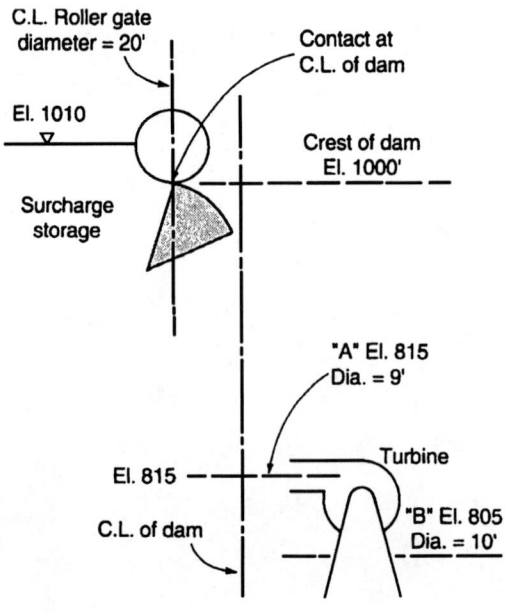

Exhibit 5.3

a. When the water surface behind the dam is at elevation 1010 ft, the horizontal hydrostatic force on the roller gate, per foot of width, and the location of the equivalent concentrated force are
   a. 6240 lb, 5.00 feet below the water surface
   b. 6240 lb, 6.67 feet below the water surface
   c. 3120 lb, 5.00 feet below the water surface
   d. 3120 lb, 6.67 feet below the water surface

b. For the same water surface elevation, the vertical hydrostatic force on the roller gate, per foot of width, and the location of the equivalent concentrated force are nearest to
   a. 4900 lb, 4.25 feet left of the roller center, downward
   b. 4900 lb, 4.25 feet left of the roller center, upward
   c. 6240 lb, 5.00 feet left of the roller center, upward
   d. 1340 lb, 7.80 feet left of the roller center, downward

The diameter of the pipe at B is 10 ft, and the pressure at B is −5 ft of water. The diameter of the pipeline at A is 9 ft. The required turbine output is 5000 horsepower for power generation. Assume a constant efficiency of 91 percent and neglect friction and entrance losses.

c. Under these conditions, the turbine discharge is most nearly
   a. 970 ft^3/sec
   b. 1000 ft^3/sec
   c. 1020 ft^3/sec
   d. Cannot be determined

d. Under these conditions, the pressure at A is most nearly
   a. 13.4 lb/in.^{2}
   b. 15.2 lb/in.^{2}
   c. 17.8 lb/in.^{2}
   d. 19.5 lb/in.^{2}

5.4 Consider the hydraulic system in Exhibit 5.4. The water surface in reservoir X behind the dam is at 300 m elevation. The outlet line, X, which leaves the dam at an elevation of 270 m, has a diameter of 60 cm and is 3000 m long. This line divides at point P into two lines of 45 cm diameter, called line Y and line Z. All three pipes are of relatively smooth steel. Line Y discharges 0.6 m^3/sec through 1500 m of pipe to tank Y when the water surface elevation in tank Y is at 150 m. Line Z is 1200 m long and 45 cm in diameter and ends at tank Z.

a. The elevation of the energy line at P is most nearly
   a. 181.4 m
   b. 179.0 m
   c. 159.6 m
   d. 121.0 m

b. The elevation of the water surface in tank Z is most nearly
   a. 101.0 m
   b. 94.7 m
   c. 88.6 m
   d. 85.5 m

El. 300 m ▽

Reservoir X

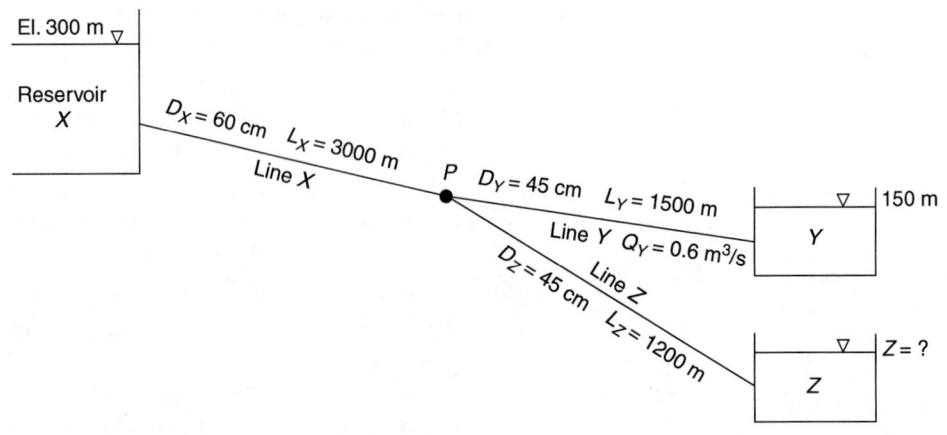

$D_X = 60$ cm $L_X = 3000$ m
Line X

P $D_Y = 45$ cm $L_Y = 1500$ m
Line Y $Q_Y = 0.6$ m³/s

▽ 150 m
Y

$D_Z = 45$ cm
Line Z
$L_Z = 1200$ m

▽ Z = ?
Z

Exhibit 5.4

5.5 A concrete-lined trapezoidal open channel with a slope of 0.34 percent has a base width of 1.5 m and side slopes of 2:1, as shown in Exhibit 5.5. This channel is designed to convey a peak discharge of 15 m³/sec. Assume $n = 0.015$.

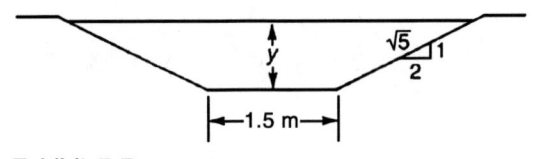

Exhibit 5.5

a. The depth of flow is most nearly
   a. 1.00 m
   b. 1.15 m
   c. 1.25 m
   d. 1.35 m

b. The flow condition, based on the Froude number, is
   a. subcritical
   b. critical
   c. supercritical
   d. not determinable for nonrectangular channels

5.6 A pump is to deliver water from one reservoir to another in which the water surface is 100 feet higher. Between the pump and the first reservoir is 2000 feet of 4-inch-diameter pipe, and 1000 feet of 6-inch-diameter pipe connects the pump and the second reservoir. Commercial steel pipe is used. The head-discharge characteristic curve for the pump is described by this table:

Discharge, gallons/min.	Head, feet
0	300
100	275
200	248
300	217
400	180
500	126

The discharge is most nearly
  a. 285 gallons/min.
  b. 295 gallons/min.
  c. 305 gallons/min.
  d. 315 gallons/min.

5.7 A power company has a suitable site for a hydroelectric pumped-storage plant near an arch dam. To satisfy downstream uses, 550 ft^3/sec must be discharged downstream from the new unit at all times. The power generated from this discharge through the power plant will be used to pump the remaining water to an elevated storage area. The water will be lifted by a turbine pump that is 93 percent efficient overall. Yield studies have indicated a firm yield of 780 ft^3/sec will be available for storage at all times.

A good storage site is available some distance away. The water will be conveyed to the storage area via a 4000 ft steel conduit (equivalent to riveted steel pipe) followed by a 5000 ft concrete-lined trapezoidal channel. Assume the channel base width and design water depth are identical. A diagram of the system is given in Exhibit 5.7.

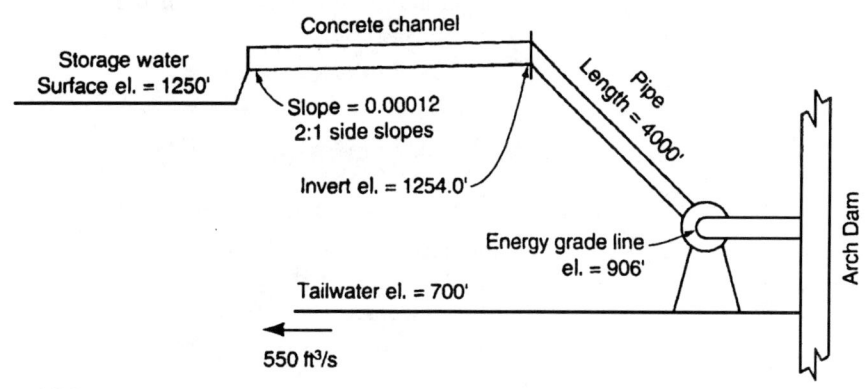

Exhibit 5.7

The most appropriate pipe diameter for use between the pump and the trapezoidal channel is
  a. 48 in.
  b. 51 in.
  c. 54 in.
  d. 60 in.

5.8 The drawing shown in Exhibit 5.8 is a cross section through the entrance of a concrete-lined irrigation canal at a low-diversion dam. The reservoir water surface is maintained at 2.0 m above the canal invert. Assume a Manning n of 0.013.

a. If the canal invert slope is 0.0008, the maximum discharge through the canal is most nearly
  a. 9.0 m^3/sec
  b. 10.5 m^3/sec
  c. 12.0 m^3/sec
  d. 13.5 m^3/sec

80 Chapter 5 Hydraulics

Exhibit 5.8

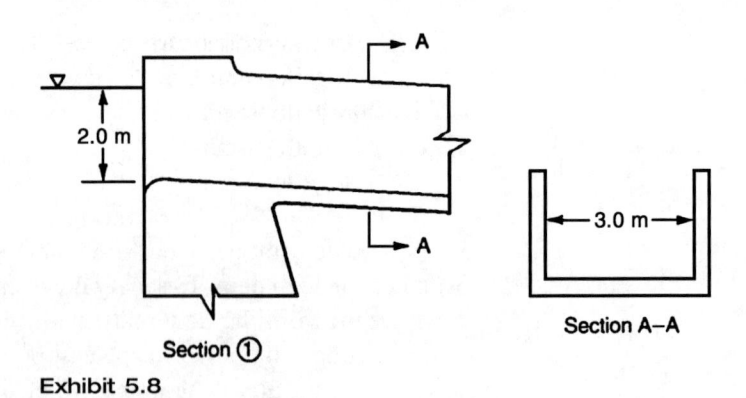

Section ①
Section A–A

b. As the invert slope is increased, the slope at which no further increase in discharge occurs is most nearly
  a. 0.0120
  b. 0.0070
  c. 0.0035
  d. 0.0015

c. The maximum discharge for the condition described in **b** is most nearly
  a. 7.0 m³/sec
  b. 14.0 m³/sec
  c. 21.0 m³/sec
  d. 28.0 m³/sec

5.9 The Foothill Ditch Company has been supplying water to a lumber company through a semicircular flume made of corrugated metal pipe (CMP) that is 12.0 feet wide. The water delivery rate has been 425 ft³/sec when the water depth was 5.0 feet. Because of certain operational and maintenance problems, the company proposes to line the flume with a 1.5-inch-thick cement mortar lining, as shown in Exhibit 5.9. Assume $n = 0.030$ for the corrugated metal pipe (CMP) and $n = 0.013$ for the mortar lining.

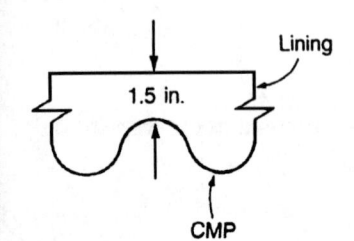

Exhibit 5.9

a. After the flume is lined with cement mortar, the discharge when the depth of flow is 5.0 ft is most nearly
  a. 425 ft³/sec
  b. 630 ft³/sec
  c. 960 ft³/sec
  d. 980 ft³/sec

b. The mean velocity in the newly lined flume is most nearly
  a. 9.5 ft³/sec
  b. 14.3 ft³/sec
  c. 21.8 ft³/sec
  d. 22.0 ft³/sec

c. The flow states before and after the lining is placed are, respectively,
  a. subcritical and subcritical
  b. subcritical and supercritical
  c. supercritical and subcritical
  d. supercritical and supercritical

5.10 A pump and distribution reservoir are to be selected to serve a new subdivision. By reference to various data and by comparison with nearby neighborhoods, the following hourly water-demand schedule is to be used for design purposes (the 24-hour period begins at midnight):

Hour	Demand, m³	Hour	Demand, m³
1	33.7	13	70.0
2	31.4	14	66.6
3	29.5	15	59.4
4	27.6	16	56.0
5	26.1	17	52.6
6	24.6	18	55.6
7	33.3	19	53.8
8	43.9	20	50.3
9	52.6	21	46.6
10	59.4	22	42.8
11	66.6	23	39.4
12	77.7	24	36.0

All water will be delivered to the reservoir by the pump.

a. If the pump operates continuously at a constant discharge, the minimum required storage volume for the reservoir is most nearly
   a. 153 m³
   b. 245 m³
   c. 890 m³
   d. 1135 m³

b. If the pump is to operate at a constant discharge only at night (10 PM to 6 AM), the required pumping rate is most nearly
   a. 48 m³/hr
   b. 96 m³/hr
   c. 142 m³/hr
   d. 192 m³/hr

c. If the pump is to operate at a constant discharge only at night (10 PM to 6 AM), the minimum required storage volume is most nearly
   a. 153 m³
   b. 245 m³
   c. 890 m³
   d. 1135 m³

5.11 The City of Excelsior is presently supplied by two pipelines from a reservoir located 47 miles away. Exhibit 5.11 is a schematic diagram of the system, and the accompanying table presents pipe data.

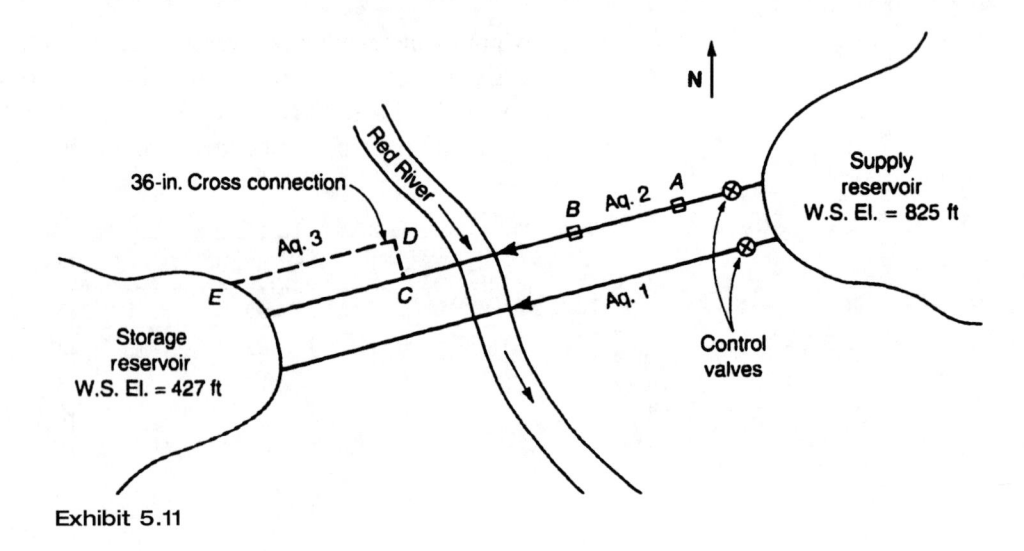

Exhibit 5.11

Exhibit 5.11a Pipe data

	Aqueduct 1	Aqueduct 2	Aqueduct 3	36-in. Cross connection
Length	47 miles	47 miles	10 miles	20 feet
Diameter	56 in.	60 in.	78 in.	36 in.
Darcy f	0.013	0.012	0.012	0.012
Lining type	Mortar	Enamel	Enamel	Enamel

To meet water demands in the year 2020, it will be necessary to construct a third pipeline. However, full capacity of the new aqueduct will not be required until the year 2020. In the interim, only a section from the west bank of the Red River to the city storage reservoir will be constructed. It will be built of 78-inch enameled steel pipe with a temporary 36-inch cross connection west of the river.

a. The discharge when the control valve is fully open in aqueduct 1 is closest to
   a. 62.5 mgd
   b. 65.0 mgd
   c. 67.5 mgd
   d. 70.0 mgd

b. The discharge when the control valve is open in the original aqueduct 2 is closest to
   a. 70 mgd
   b. 75 mgd
   c. 80 mgd
   d. 85 mgd

c. Now consider aqueduct 2 after the section called aqueduct 3 has been constructed. The sections of these two pipes west of the river could, for analysis purposes, be replaced with an "equivalent pipe." Compared with the two individual parallel pipe segments, the equivalent pipe would most nearly have
   a. a head loss equal to the sum of the head losses in the original pipe sections and a discharge that matches the flow in the larger pipe
   b. a discharge equal to the sum of the discharges in the two original pipe sections but the same head loss as the original pipes
   c. a discharge equal to the sum of the discharges in the two original pipe sections but a lower overall head loss
   d. a smaller head loss than that of the original combination

d. Determine the discharge in aqueduct 2 from the supply reservoir after the new section, aqueduct 3, has been built. The flow is nearest
   a. 85 mgd
   b. 90 mgd
   c. 95 mgd
   d. 100 mgd

e. The discharge in aqueduct 3 is most nearly
   a. 40 mgd
   b. 50 mgd
   c. 60 mgd
   d. 70 mgd

f. The head loss in aqueduct 3 is most nearly
   a. 0.2 foot
   b. 10.0 feet
   c. 390 feet
   d. 398 feet

g. The head loss from friction in the cross connection is most nearly
   a. 0.2 foot
   b. 10.0 feet
   c. 390 feet
   d. 398 feet

h. Some field tests were conducted after aqueduct 3 had been in service for several months. Pressure gage readings were taken at four points when the flow in aqueduct 2 was 90 mgd (V_2 = 7.09 ft/sec) and at the same time the flow in aqueduct 3 was 60 mgd (V_3 = 2.80 ft/sec). Points A and B are separated by 113,000 ft, and points B and C are separated by 5700 ft.

   Based on the data, the actual Darcy friction factor, f, in aqueduct 3 is closest to
   a. 0.0120
   b. 0.0123
   c. 0.0126
   d. 0.0130

Exhibit 5.11b Pressure gage readings

Point	Elev. of Gage, ft	Gage Reading, lb/in.^{2}
A	215.0	207.0
B	49.0	179.0
D	46.7	170.0
E	420.0	3.0

i. Based on the given data, the actual value of the coefficient C_{HW} in the Hazen-Williams formula for aqueduct 2 is nearest to
   a. 128
   b. 132
   c. 136
   d. 140

j. Using the given data, the overall head loss in the cross connection is most nearly
   a. 0.2 foot
   b. 6.0 feet
   c. 9.0 feet
   d. 12.0 feet

SOLUTIONS

5.1 In this problem, local head losses will be neglected, and $f = 0.021$. The given length of the pipe is 5.1 mi × 5280 ft/mi = 26,900 ft. The head loss is equal to the difference in elevations of the pipe ends, of $h_L = 280 - 100 = 180$ ft. The present demand is

$$1 \text{ mgd} = (10^6 \text{ gal})\left(\frac{1 \text{ ft}^3}{7.48 \text{ gal}}\right)\frac{1}{60 \times 60 \times 24 \text{ sec/day}} = 1.55 \text{ ft}^3/\text{sec}$$

a. c. The minimum pipe diameter that could be used is found from the Darcy-Weisbach equation (5.10) in the text:

$$h_L = f\frac{L}{D}\frac{V^2}{2g} = f\frac{L}{D}\frac{Q^2}{2g\left(\frac{\pi}{4}D^2\right)^2}$$

Here,

$$D^5 = \frac{8fLQ^2}{h_L g\pi^2} = \frac{8(0.021)(26,900)(1.55)^2}{(180.0)(32.2)\pi^2}$$
$$D^5 = 0.190 \text{ ft}^5$$
$$D = 0.717 \text{ ft} = 8.6 \text{ in}$$

The next larger commercially available size should be used, which in this case is 10 in. nominal inside diameter cast-iron pipe.

b. **b.** The friction factor f is a function of Reynolds number and relative roughness; these two factors will be determined, and then f can be read from the Moody diagram, Fig. 5.2, in *Civil Engineering PE License Review*. In this calculation, the diameter will be assumed to be 10 in = 0.833 ft, and

$$V = \frac{Q}{A} = \frac{1.55}{\frac{\pi}{4}(0.833)^2} = 2.84 \frac{\text{ft}}{\text{sec}}$$

If 60 °F is assumed as the water temperature, then the kinematic viscosity is $v = 1.2 \times 10^{-5}$ ft^2/sec and

$$\text{Re} = \frac{VD}{v} = \frac{(2.84)(0.833)}{1.2 \times 10^{-5}} = 2.0 \times 10^5$$

Use Table 5.1 in *Civil Engineering PE License Review* to find $\varepsilon = 0.0102$ in. Then $\varepsilon/D = 0.0102/10 = 0.001$. For these values, the Moody diagram gives $f = 0.021$ or slightly higher. The assumed value is appropriate.

c. **a.** If the Manning equation [Eq. (5.29)] were used in this problem, then

$$V = \frac{1.49}{n} R^{2/3} S_o^{1/2}$$

in which $S_o = h_L/L$ and $R = D/4$. If this equation is solved for h_L and arranged into the same form as the Darcy equation, then

$$h_L = f \frac{L}{D} \frac{V^2}{2g} = \left(\frac{n^2 2g 4^{4/3}}{1.49^2 D^{1/3}}\right) \frac{L}{D} \frac{V^2}{2g}$$

and

$$n^2 = \frac{fD^{1/3}}{184} = \frac{(0.021)(0.833)^{1/3}}{184} = 0.000107$$

Hence, $n = 0.010$. A check of Table 5.4 shows that this value is correct for clean, new cast iron.

5.2 b. In this solution, the subscripts A and B refer to the 3 m and 15 m pipes, respectively. The subscript 25 refers to the 25 cm pipe, and the subscript 15 refers to the 15 cm pipes.

Neglecting minor losses, one can write a pair of energy equations, one between the reservoir and outlet A and one between the reservoir and outlet B, and also one continuity equation at the intersection of the 25-cm and 15-cm pipes. These equations are

$$40 - f\frac{L_{25}}{D_{25}} \frac{V_{25}^2}{2g} - f\frac{L_A}{D_{15}} \frac{V_A^2}{2g} = \frac{V_A^2}{2g} + 20 \tag{a}$$

$$40 - f\frac{L_{25}}{D_{25}} \frac{V_{25}^2}{2g} - f\frac{L_B}{D_{15}} \frac{V_B^2}{2g} = \frac{V_B^2}{2g} + 20 \tag{b}$$

$$A_{15}(V_A + V_B) = A_{25} V_{25} \tag{c}$$

From (a) and (b),

$$\left(f\frac{L_A}{D_{15}}+1\right)V_A^2 = \left(f\frac{L_B}{D_{15}}+1\right)V_B^2$$

$$\frac{V_A}{V_B} = \left(\frac{fL_B+D_{15}}{fL_A+D_{15}}\right)^{1/2} = \left(\frac{(0.02)(15)+0.15}{(0.02)(3)+0.15}\right)^{1/2} = 1.464$$

Now the continuity equation (c) yields

$$V_{25} = \frac{A_{15}}{A_{25}}(V_A+V_B) = \left(\frac{15}{25}\right)^2(2.464\,V_B) = 0.887V_B$$

and substitution of these results into (b) gives

$$20(2g) = \left\{f\left[\frac{L_{25}}{D_{25}}(0.887)^2 + \frac{L_B}{D_{15}}\right]+1\right\}V_B^2$$

or

$$V_B^2 = \frac{(20)(2)(9.81)}{0.02\left[\frac{30}{0.25}(0.887)^2 + \frac{15}{0.15}\right]+1} = 80.3$$

Hence, $V_B = 8.96$ m/sec, $V_{25} = 0.887V_B = 7.95$ m/sec, and

$$Q = A_{25}V_{25} = \frac{\pi}{4}(0.25)^2(7.95) = 0.390 \text{ m}^3\text{/sec}$$

5.3

a. d. Exhibit 5.3a presents a diagram of the forces on the roller gate. This diagram shows a partial free-body diagram of the forces on the chunk of fluid bounded by *CDE*. The horizontal force on *CD* is

$$H = F_H = \frac{\gamma 10^2}{2} = 50\gamma = 50(62.4) = 3120 \text{ lb/ft}$$

in which $\gamma = 62.4$ lb/ft^3 is the unit weight of water and this force is identical to the horizontal force acting on the curved surface *CE* itself. This force is the integral of the triangular pressure distribution shown in the figure, and *H* therefore acts a distance $\frac{2}{3}(10) = 6.67$ ft below the water surface at elevation 1010 ft.

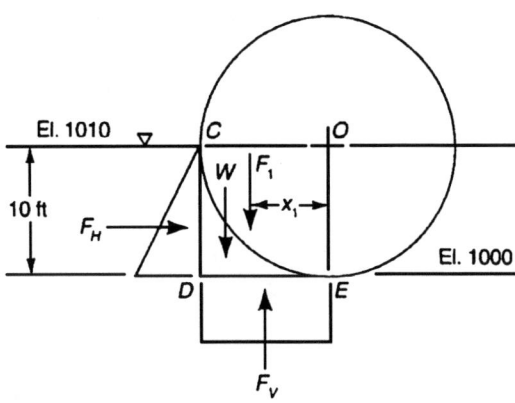

Exhibit 5.3a

b. **b.** The determination of the vertical force can be computed in either of two ways. From Exhibit 5.3a, vertically the force between the gate and the water is the difference between a vertical force F_V acting upward on DE and the weight W of the water in sector CDE acting downward. The net vertical force per foot width is

$$V = F_V - W = 10\gamma(10) - \gamma\left[10^2 - \frac{\pi}{4}10^2\right] = \frac{\pi}{4}10^2\gamma = 4900 \text{ lb}$$

When one recognizes this vertical force as simply the reaction F_1 to the weight of water that could be contained in the quarter-circle OCE, then the location of F_1 is at the centroid of the quarter-circle, which is at

$$x_1 = \frac{4R}{3\pi} = \frac{4(10)}{3\pi} = 4.24 \text{ ft}$$

c. **c.** The fluid power that can be created is $Q\gamma E_m$, in which Q is the discharge and E_m is the change in head across the turbine, but the turbine output is reduced by the efficiency factor $\eta = 0.91$, so the actual output will be

$$P = 5000 = \frac{\eta Q \gamma E_m}{550}$$

Since the required power output is to be a constant, the head-discharge curve will assume the form $QE_m = $ constant, which is hyperbolic. Thus

$$QE_m = \frac{550P}{\eta\gamma} = \frac{550(5000)}{0.91(62.4)} = 48,400$$

Write the Bernoulli equation between the reservoir water surface and point B:

$$850 = 805 + \frac{p_B}{\gamma} + \frac{V_B^2}{2g} + E_m$$

The velocity at B is

$$V_B = \frac{Q}{A_B} = \frac{Q}{\frac{\pi}{4}(10)^2} = \frac{Q}{25\pi}$$

Since the pressure head at B is specified as $p_B/\gamma = -5$ ft, the Bernoulli equation becomes

$$50 = \frac{1}{2(32.2)}\left(\frac{Q}{25\pi}\right)^2 + E_m = \frac{Q^2}{397,000} + E_m$$

Solving this equation for Q will establish the discharge. Then V_A and p_A will be found. The equation $E_m = 48,400/Q$ can be inserted in this last equation, and Q can then be found by successive trials. Clearly E_m will be less than 50 ft.

Trial	Q, ft³/sec	$\frac{Q^2}{397,000} + E_m \stackrel{?}{=} 50.0$
1	1000	2.5 + 48.4 = 50.9 (High)
2	1050	2.78 + 46.1 = 48.9 (Low)
3	1020	2.62 + 47.5 = 50.1 (OK) $Q \approx 1020$ ft³/sec

d. a. From continuity

$$V_A = \frac{Q}{A_A} = \frac{1020}{\frac{\pi}{4}(9)^2} = 16.0 \text{ ft/sec}$$

Writing the Bernoulli equation between point A and the reservoir water surface now gives

$$850 = 815 + \frac{p_A}{\gamma} + \frac{V_A^2}{2g} = 815 + \frac{p_A}{\gamma} + \frac{(16.0)^2}{2(32.2)}$$

Hence,

$$\frac{p_A}{\gamma} = +31.0 \text{ ft} \quad \text{or} \quad p_A = 31.0\left(\frac{62.4}{144}\right) = 13.4 \text{ lb/in.}^2$$

5.4 As a preliminary remark, it is not totally clear whether to consider or neglect local losses in this problem. Certainly one velocity head will be lost on entering tank Y or Z, but the exit condition on leaving reservoir X is not stated, and in any event, local losses at P are difficult to treat. To be consistent, neglect local losses in this problem.

Since all pipes are of relatively smooth steel, from Table 5.1 in *Civil Engineering PE License Review,* the roughness ε will be chosen as 0.046 mm. Hence, $\varepsilon/D = 0.000077$ for the 60 cm pipe, and $\varepsilon/D = 0.0001$ for the other pipes.

a. **a.** Since the water surface elevation in tank Y and the discharge to tank Y are known,

$$V_Y = \frac{Q_Y}{A_Y} = \frac{0.6}{\frac{\pi}{4}(0.45)^2} = 3.77 \text{ m/sec}$$

and the pipe Reynolds number, using a nominal kinematic viscosity for water of $\nu = 10^{-6}$ m²/sec, is

$$\text{Re}_Y = \frac{VD}{\nu} = \frac{3.77(0.45)}{10^{-6}} = 1.7(10^6)$$

The Moody diagram, Fig. 5.2 in *Civil Engineering PE License Review,* now gives $f_Y = 0.013$. Between tank Y and point P, the energy line yields

$$h_P = 150 + f_Y \left(\frac{L}{D}\right)_Y \frac{V_Y^2}{2g} = 150 + 0.013 \left(\frac{1500}{0.45}\right) \frac{(3.77)^2}{2(9.81)}$$

or

$$h_P = 18.4 \text{ m}$$

b. **b.** Between reservoir X and point P the energy equation is

$$300 - f_X \left(\frac{L}{D}\right)_X \frac{V_X^2}{2g} = h_P = 181.4$$

Assuming the flow to be in the region of complete turbulence in the Moody diagram leads to $f_X = 0.0115$. Thus,

$$0.0115 \left(\frac{3000}{0.6}\right) \frac{V_X^2}{2(9.81)} = 300 - 181.4 = 118.6 \quad \text{and} \quad V_X = 6.36 \text{ m/sec}$$

A computation of the Reynolds number

$$\text{Re}_X = \frac{V_X D}{\nu} = 3.8(10^6)$$

shows that the friction factor should be adjusted to $f_X = 0.012$. Using this value in the energy equation yields

$$V_X = 6.22 \text{ m/sec}$$

Thus,

$$Q_X = V_X A_X = 6.22 \left(\frac{\pi}{4}\right)(0.6)^2 = 1.76 \text{ m}^3\text{/sec}$$

Applying continuity at point P,

$$Q_Z = Q_X - Q_Y = 1.76 - 0.60 = 1.16 \text{ m}^3/\text{sec}$$

Consequently,

$$\text{Re}_Z = 3.3 \ (10^6) \quad \text{and} \quad f_Z = 0.012$$

Finally, the energy equation between point P and tank Z gives

$$h_P - f_Z \left(\frac{L}{D}\right)_Z \frac{V_Z^2}{2g} = Z$$

or

$$Z = 181.4 - 0.012 \left(\frac{1200}{0.45}\right) \frac{(7.29)^2}{2(9.81)} = 94.7 \text{ m}$$

5.5

a. c. The depth of flow y is found from the Manning equation [Eq. (5.29)]:

$$Q = \frac{1}{n} A R^{2/3} S_o^{1/2}$$

For this channel,

$$15 = \frac{1}{0.015} A R^{2/3} (0.0034)^{1/2} \quad \text{or} \quad AR^{2/3} = 3.86$$

For the given data, $A = (1.5 + 2y)y$, $P = 1.5 + 2\sqrt{5}y$, and $R = A/P$. Although other methods could be used to solve this equation for y, the use of a simple table of trials is less error prone and is recommended for exam problems. A sample solution follows in which one first seeks to straddle the actual answer with preliminary trials and then narrows the interval to obtain a sufficiently accurate result:

Trial	y	A	P	R	$AR^{2/3} \stackrel{?}{=} 3.86$
1	1.0	3.50	5.97	0.59	2.45 (Low)
2	1.3	5.33	7.31	0.73	4.31 (High)
3	1.2	4.68	6.87	0.68	3.62 (Low)

Now enough is known about the depth, y, to select the correct result, but one could continue to refine the result to find eventually that y is between 1.23 m and 1.24 m.

b. c. The Froude number, and its square, can be written in several forms. One of the most versatile is in terms of the hydraulic depth $D = A/b$, in which A = flow cross-sectional area and b = channel width at the water surface:

$$\text{Fr}^2 = \frac{V^2}{gD} = \frac{Q^2}{gA^3/b} = \frac{Q^2 b}{gA^2}$$

This form applies equally well to nonrectangular channels. In particular, when $y = 1.25$ m, $b = 1.5 + 4(1.25) = 6.5$ m, $A = [1.5 + 2(1.25)](1.25) = 5.0$ m^2, and

$$\text{Fr}^2 = \frac{(15)^2(6.5)}{(9.81)(5.0)^3} = 1.2 > 1$$

If one simply uses the depth y in place of D, one finds $\text{Fr}^2 < 1$, which is incorrect here.

5.6 d. Two initial comments are appropriate here: (1) Although several types of pipe friction relations may normally be used in such problems, the form of the data given here makes the use of the Darcy-Weisbach formula most appropriate; (2) A plot of head E_p versus discharge Q, as shown in Exhibit 5.6, may be more useful than the table.

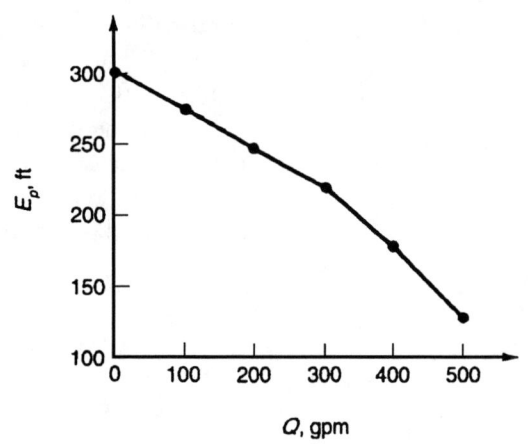

Exhibit 5.6

A diagram of the system, introducing some notation, is shown in Exhibit 5.6a.

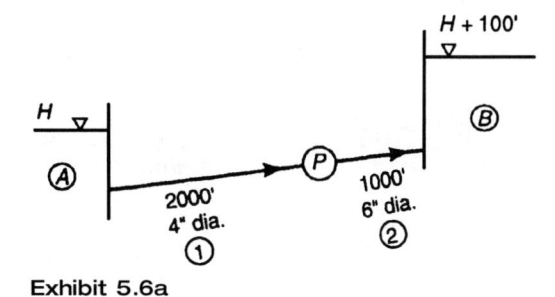

Exhibit 5.6a

The pipes labeled 1 and 2 are long enough so that local losses may be neglected. First, the energy equation is written between reservoirs A and B to obtain

$$H - f_1 \frac{L_1}{D_1} \frac{V_1^2}{2g} + E_p - f_2 \frac{L_2}{D_2} \frac{V_2^2}{2g} = H + 100$$

From continuity

$$Q = A_1 V_1 = A_2 V_2$$

Hence,

$$\left(\frac{\pi}{4}\right) 4^2 V_1 = \left(\frac{\pi}{4}\right) 6^2 V_2 \quad \text{or} \quad V_1 = \frac{9}{4} V_2$$

The energy equation becomes

$$E_p - 100 = \left[f_1 \frac{L_1}{D_1} \left(\frac{9}{4}\right)^2 + f_2 \frac{L_2}{D_2} \right] \frac{V_2^2}{2g}$$

Once f_1, f_2, and E_p are estimated, V_2 may be computed. Then additional calculations will be made to check the correctness of the estimates. The process continues until the estimates and other calculations agree sufficiently well.

For commercial steel pipe, Table 5.1 in *Civil Engineering PE License Review* gives $\varepsilon = 0.0018$ in. If the pipe Reynolds numbers are sufficiently high that they are in the wholly rough region of the Moody diagram, Fig. 6.2 in the text, then f is determined only by ε/D:

$$\left(\frac{\varepsilon}{D}\right)_1 = \frac{0.0018}{4} = 0.00045 \quad \text{implies} \quad f_1 = 0.0165$$

$$\left(\frac{\varepsilon}{D}\right)_2 = \frac{0.0018}{6} = 0.00030 \quad \text{implies} \quad f_2 = 0.0150$$

Clearly one must have $E_p > 100$ ft for the discharge to be positive. To get the calculations started, estimate $E_p = 200$ ft. Using $g = 32.2$ ft/sec^2, one has

$$200 - 100 = \left[(0.0165) \frac{2000}{4/12} \left(\frac{9}{4}\right)^2 + (0.0150) \frac{100}{6/12} \right] \frac{V_2^2}{2g} \quad \text{or} \quad V_2 = 3.48 \text{ ft/sec}$$

This implies

$$Q = A_2 V_2 = (3.48) \frac{\pi}{4} \left(\frac{6}{12}\right)^2 = 0.684 \frac{\text{ft}^3}{\text{sec}} = 307 \text{ gallons/min}$$

since 1 ft³/s = 449 gallons/min. Now check f_1, f_2, and E_p. For $Q = 307$ gallons/min, $E_p = 214$ ft from Exhibit 5.6. For pipe 1,

$$\text{Re}_1 = \frac{V_1 D_1}{\nu} \approx \frac{\frac{9}{4}(3.48)\frac{4}{12}}{10^{-5}} = 2.6 \times 10^5 \quad \text{implies} \quad f_1 = 0.018$$

For pipe 2,

$$\text{Re}_2 = \frac{V_2 D_2}{\nu} \approx \frac{(3.48)\frac{6}{12}}{10^{-5}} = 1.7 \times 10^5 \quad \text{implies} \quad f_2 = 0.018$$

Now recompute V_2 with these values:

$$214 - 100 = \left[(0.018)\frac{2000}{4/12} \left(\frac{9}{4}\right)^2 + (0.018)\frac{1000}{6/12} \right] \frac{V_2^2}{2g}$$

or $V_2 = 3.55$ ft/s, which leads to $Q = 313$ gallons/min. An additional cycle of computations would confirm that this set of values is sufficiently accurate.

5.7 b. This solution is based on the assumption that the turbine and pump at the head of the pipe produce and use, respectively, just enough power to deliver the water to the open channel if the most efficient pipe size is chosen (that is, the smallest allowable size is selected). To begin the solution of the problem, construct an energy-line diagram for the pipe. Exhibit 5.7a presents this diagram with known information shown. The elevation of the energy line at the transition point from the pipe to the open channel is not initially known and is called W in the figure. (Local losses are ignored in this diagram.) Since W is needed now, one must calculate it. The trapezoidal channel section is lined with concrete, so choose $n = 0.013$. The channel has 2:1 side slopes, and a diagram of the channel cross section is shown in Exhibit 5.7b. Manning's equation [Eq. (5.29)] is

$$Q = \frac{1.49}{n} A R^{2/3} S_o^{1/2}$$

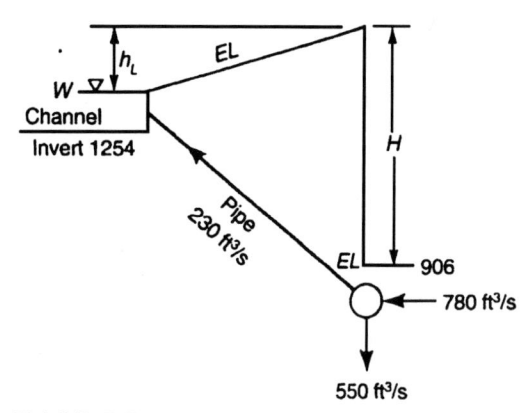

Exhibit 5.7a

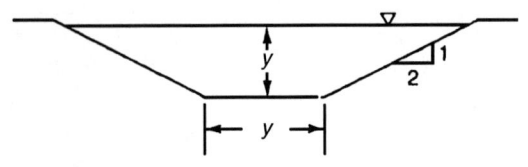

Exhibit 5.7b

With the present data

$$230 = \frac{1.49}{0.013} AR^{2/3}(0.00012)^{1/2}$$

The area A, wetted perimeter P, and hydraulic radius R for this channel are $A = 3y^2$, $P = y + 2\sqrt{5}y$, and $R = A/P$. Hence,

$$AR^{2/3} = 3y^2 \left[\frac{3y^2}{(1+2\sqrt{5})y}\right]^{2/3} = 183.2$$

By direct solution, $y^{8/3} = 91.1$ and $y = 5.43$ ft. The open channel trapezoidal cross-sectional area is $A = 88.5$ ft^2, the velocity is $V = Q/A = 2.60$ ft/sec, and the elevation of the energy line above the channel invert is $E = y + V^2/2g = 5.43 + 0.10 = 5.5$ ft. Thus, $W = 1254.0 + 5.5$ or $W = 1259.5$ ft.

By assumption, the power generated and consumed are the same, or $Q_1 \gamma E_T \eta = Q_2 \gamma H$, in which $Q_1 = 550$ ft^3/sec, $Q_2 = 230$ ft^3/sec, $\gamma = 62.4$ lb/ft^3, E_T = energy extracted from the water by the turbine in ft-lb/lb, H = gain in energy across the pump, and η = efficiency of the turbine-pump combination = 0.93. Here, $E_T = 906 - 700 = 206$ ft, leading to

$$H = \frac{Q_1}{Q_2} E_T \eta = \frac{550}{230}(206)(0.93) = 458 \text{ ft}$$

From the energy-line diagram, $1259.5 + h_L = 906 + H$, or

$$104.5 = h_L = f\frac{L}{D}\frac{V^2}{2g} \quad \text{with} \quad V = \frac{Q}{\frac{\pi}{4}D^2}$$

Upon combining these expressions,

$$\frac{D^5}{f} = \frac{16Q^2 L}{2g\pi^2 h_L} = \frac{16(230)^2(4000)}{2(32.2)\pi^2(104.5)} = 51{,}000$$

Now this equation must be solved for D; this will be done via trial solutions that are recorded in a table. Since the pipe and the discharge are large, the Reynolds number is expected to be large, but this must be checked at the end of the solution. If the Reynolds number is sufficiently large, then the friction factor is essentially a function only of the relative roughness ε/D. One uncertainty in this problem is the value of ε, since Table 5.1 presents a range of roughness values for riveted steel. From this range, a mean value of 0.1 inch will be used.

Trial	D, ft	ε/D	f	$D^5/f \stackrel{?}{=} 51{,}000$
1	5.00	0.00167	0.022	142,000 (High)
2	4.00	0.00208	0.0235	43,600 (Low)
3	4.25	0.00196	0.0235	59,000 (High)

The solution is between the last two trial solutions, and f is constant in this narrow range, so $D^5 = 51{,}000f = 51{,}000(0.0235) = 1200$ and $D = 4.13$ ft. The velocity of flow in a pipe of this size is

$$V = \frac{Q}{A} = \frac{230}{\frac{\pi}{4}(4.13)^2} = 17.2 \text{ ft/sec}$$

The approximate Reynolds number is

$$\text{Re} = \frac{VD}{\nu} = \frac{(17.2)(4.13)}{10^{-5}} = 7 \times 10^6$$

which is sufficiently large. One should choose the next commercially available pipe size above $D = 4.13$ ft.

The solution that has just been presented is both correct and thorough, but it is rather long. It is worth noting that, if one ignored entirely the open-channel calculations and simply used W as 1254 ft rather than 1259.5 ft, the final computed pipe diameter would be $D = 4.08$ ft.

In either case, the next larger commercially available pipe size should be chosen. In this range, the increment in pipe diameter is 3 inches, so one should choose 4.25 ft or 51 in. for the diameter.

5.8

a. b. At the invert in the canal entrance, the specific energy is

$$E = 2.00 = y + \frac{V^2}{2g}$$

and the velocity is given by the Manning formula in metric units as

$$V = \frac{1}{n} R^{2/3} S_o^{1/2}$$

with the hydraulic radius

$$R = \frac{A}{P} = \frac{3y}{3+2y}$$

Hence,

$$y + \frac{1}{2(9.81)}\left(\frac{1}{0.013}\right)^2 \left(\frac{3y}{3+2y}\right)^{4/3}(0.0008) = 2.00$$

Since one must have $y < 2.00$ m, one can begin a successive trial solution of the equation with a good estimate. One has

$$E(y) = y + 0.241\left(\frac{3y}{3+2y}\right)^{4/3} = 2.00$$

If $y = 1.80$ m, then $E(y) = 1.80 + 0.184 = 1.984$ m. Since the larger term in this sum is the depth, simply add 0.016 to y for the next trial. Thus, $E(1.816) = 1.816 + 0.185 = 2.00$ m. Hence, $y = 1.816$ m. The discharge is then given by

$$Q = \frac{1}{n}AR^{2/3}S_o^{1/2} \quad \text{with} \quad R = \frac{3(1.816)}{3+2(1.816)} = 0.821 \text{ m}$$

$$Q = \frac{1}{0.013}[3(1.816)](0.821)^{2/3}(0.0008)^{1/2} = 10.4 \text{ m}^3/\text{sec}$$

b. c. The maximum discharge first occurs when the slope increases sufficiently to become the critical slope S_c; any further steepening beyond S_c will not increase the discharge any more. The corresponding depth of flow is the critical depth y_c, which is, for rectangular channels only,

$$y_c = \frac{2}{3}E = \frac{2}{3}(2.00) = 1.333 \text{ m}$$

Thus,

$$y_c + \frac{V_c^2}{2g} = 2.00 \quad \text{and} \quad V_c = [(2.00-1.333)(2)(9.81)]^{1/2} = 3.62 \text{ m/sec}$$

Also,

$$V_c = \frac{1}{n}R_c^{2/3}S_c^{1/2}$$

Now compute R_c and S_c in turn:

$$R_c = \frac{3(1.333)}{3+2(1.333)} = 0.706 \text{ m}$$

$$S_c = \frac{[3.62(0.013)]^2}{(0.706)^{4/3}} = 0.00353$$

c. b. A direct application of continuity gives

$$Q = V_c A_c = 3.62\,[3(1.333)] = 14.47 \text{ m}^3/\text{sec}$$

5.9 Although this problem could be solved by using specialized tables from books such as *Handbook of Hydraulics* by King and Brater (McGraw-Hill), the solution presented here will not do so. In solving this problem, one must use expressions for the area A, wetted perimeter P, hydraulic radius R, and top width b for a circular pipe flowing partially full. These expressions can be found in most texts on open-channel hydraulics and are reproduced here, using the notation defined in Exhibit 5.9a.

$$\sin \theta_1 = \frac{D/2 - y}{D/2} = 1 - \frac{2y}{D}$$

$$2 + \theta_1 = \pi$$

$$A = (\theta - \sin \theta)D^2/8$$

$$P = \theta D/2$$

$$R = \left(1 - \frac{\sin \theta}{\theta}\right)\frac{D}{4}$$

$$b = 2[y(D-y)]^{1/2}$$

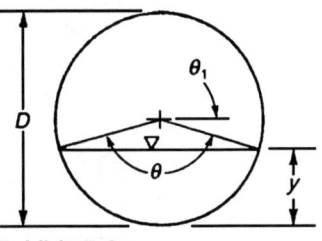

Exhibit 5.9a

a. c. The discharge in the new, mortar-lined flume will be found by using the Manning equation, but first one must use the Manning equation with data from the corrugated metal pipe (CMP) flume to establish the slope of the flume. For the CMP flume, when $y = 5$ ft and $D = 12$ ft,

$$\sin \theta_1 = 1 - \frac{2(5)}{12}$$
$$\theta_1 = 9.59°$$
$$\theta = 160.8° = 2.81 \text{ rad}$$

$$A = (2.81 - \sin 160.8°)(12)^2/8 = 44.7 \text{ ft}^2$$
$$P = 2.81(12)/2 = 16.86 \text{ ft}$$
$$R = A/P = 2.65 \text{ ft}$$

The Manning equation is

$$Q = \frac{1.49}{n} A R^{2/3} S_o^{1/2}$$

Inserting the known data gives

$$425 = \frac{1.49}{0.030}(44.7)(2.65)^{2/3} S_o^{1/2}$$

which yields $S_o^{1/2} = 0.1000$, $S_o = 0.0100$. The new diameter is

$$D = 12 - 2\left(\frac{1.5}{12}\right) = 11.75 \text{ ft}$$

which leads to the following computations for the new discharge:

$$\sin \theta_1 = 1 - \frac{2(5)}{11.75}$$
$$\theta_1 = 8.57°$$
$$\theta = 162.9° = 2.84 \text{ rad}$$

$A = (2.84 - \sin 162.9°)(11.75)^2/8 = 43.9 \text{ ft}^2$
$P = 2.84 (11.75)/2 = 16.69 \text{ ft}$
$R = A/P = 2.63 \text{ ft}$

$$Q = \frac{1.49}{0.013}(43.9)(2.63)^{2/3}(0.1000)$$

Therefore, $Q = 959 \text{ ft}^3/s$ in the new lined flume.

b.   c.   The new velocity is

$$V = \frac{Q}{A} = \frac{959}{43.9} = 21.8 \text{ ft/sec}$$

c.   b.   In general, the squared Froude number is

$$Fr^2 = \frac{Q^2 b}{gA^3}$$

Before lining, $b = 2[5(12 - 5)]^{1/2} = 11.83 \text{ ft}$, and

$$Fr^2 = \frac{(425)^2(11.83)}{(32.2)(44.7)^3} = 0.743$$

Subcritical flow exists in the CMP since $Fr^2 < 1$. After lining, $b = 2[5(11.75 - 5)]^{1/2} = 11.62 \text{ ft}$ and

$$Fr^2 = \frac{(959)^2(11.62)}{(32.2)(43.9)^3} = 3.92$$

Supercritical flow exists in the mortar-lined flume since $Fr^2 > 1$.

5.10   First, find the daily demand D. Call the hourly demand of discharge Q_i (that is, $Q_2 = 31.4 \text{ m}^3$). The daily demand is the sum of the 24 hourly demands, or

$$D = \sum_{i=1}^{24} Q_i = 1135.4 \text{ m}^3$$

a.   a.   The average discharge and pumping rate for the 24 hours is

$$\overline{Q} = \frac{D}{24} = 47.3 \text{ m}^3/\text{hr}$$

The minimum storage volume can now be found by summing the positive hourly differences between demand and supply. In this problem, $Q_i > \overline{Q}$ for $i = 9$ to $i = 20$, as can be seen by inspecting the original demand table. The minimum volume is

$$V = \sum_{i=9}^{20}(Q_i - \overline{Q}) = \sum_{i=9}^{20} Q_i - 12\overline{Q} = 152.9 \text{ m}^3$$

b.   c.   If all the pumping must be done in 8 hours, then the constant pumping rate must be

$$D/8 = 1135.4/8 = 142 \text{ m}^3/\text{hr}$$

One should note that this amount greatly exceeds any of the hourly demands.

c.   c.   The required storage is now determined by the need to store all of the water that will be used in the 16 hours when no pumping occurs. This volume is

$$V = \sum_{i=7}^{22} Q_i = 887.1 \text{ m}^3$$

5.11

a.   c.   The great length of each aqueduct line allows one to neglect all local losses when the control valve is fully open. Thus, the difference in reservoir elevations H is equal to the frictional head lost in the pipe:

$$H = f \frac{L}{D}\frac{V^2}{2g} = 825 - 427 = 398 \text{ ft}$$

Continuity may thus be written as

$$Q = AV = \frac{\pi}{4}D^2 \left[2g \frac{H}{f}\frac{D}{L}\right]^{1/2}$$

For aqueduct 1,

$$Q_1 = \frac{\pi}{4}\left(\frac{56}{12}\right)^2 \left[2(32.2)\frac{(398)}{(0.013)}\frac{(56/12)}{(47)(5280)}\right]^{1/2} = 104 \text{ ft}^3/\text{sec}$$

$$Q_1 = \left(104 \frac{\text{ft}^3}{\text{s}}\right)\left(7.48 \frac{\text{gal}}{\text{ft}^3}\right)\left(3600 \times 24 \frac{\text{s}}{\text{day}}\right)(10^{-6}) = 67.2 \text{ mgd}$$

b.   d.   Only the factors f and D change from the first to the second aqueduct. By direct proportion from the equations in the solution to part **a**,

$$Q_2 = Q_1 \left(\frac{D_2}{D_1}\right)^2 \left[\frac{f_1}{f_2}\frac{D_2}{D_1}\right]^{1/2} = (67.2)\left(\frac{60}{56}\right)^2 \left[\frac{0.013}{0.012}\frac{60}{56}\right]^{1/2} = 83.1 \text{ mgd}$$

c. **b.** Let the head loss and discharge of the equivalent pipe be h_L and Q, respectively. Let subscripts 1 and 2 denote properties of the two original parallel pipes. The equivalent pipe will function in the same way as the combination of the two original pipes only when the following two criteria are met:

$$h_L = h_{L_1} = h_{L_2}$$

$$Q = Q_1 + Q_2$$

d. **b.** It is useful to digress a bit and develop the equivalent pipe relations more thoroughly. The relation between head loss and discharge for each of the original parallel pipes is

$$h_{L_i} = K_i Q_i^n$$

For example, for the Darcy-Weisbach head loss formula,

$$K = f \frac{L}{D} \frac{1}{A^2} \frac{1}{2g} \quad \text{and} \quad n = 2$$

For the equivalent pipe,

$$KQ^n = K_1 Q_1^n = K_2 Q_2^n \quad \text{and} \quad Q = Q_1 + Q_2$$

Therefore, if K_1 and K_2 can be computed, then K is also known and can be found from

$$\left(\frac{1}{K}\right)^{1/n} = \left(\frac{1}{K_1}\right)^{1/n} + \left(\frac{1}{K_2}\right)^{1/n}$$

Now part **d** can be answered. (The effect of the cross-connection pipe is minor and can be omitted here.)

For the parallel section of pipe 2,

$$K_2 = f_2 \frac{L_2}{D_2} \frac{1}{A_2^2} \frac{1}{2g} = (0.012) \frac{10(5280)}{60/12} \frac{1}{\left[\frac{\pi}{4}\left(\frac{60}{12}\right)^2\right]^2} \frac{1}{2(32.2)} = 0.00510$$

Similarly,

$$K_3 = f_3 \frac{L_3}{D_3} \frac{1}{A_3^2} \frac{1}{2g} = (0.012) \frac{10(5280)}{78/12} \frac{1}{\left[\frac{\pi}{4}\left(\frac{78}{12}\right)^2\right]^2} \frac{1}{2(32.2)} = 0.00137$$

Using

$$\left(\frac{1}{K}\right)^{1/2} = \left(\frac{1}{0.00510}\right)^{1/2} + \left(\frac{1}{0.00137}\right)^{1/2}$$

one obtains $K = 0.000594$.

Between the supply and storage reservoirs

$$H = KQ^2 + (0.012)\frac{37(5280)}{60/12}\frac{1}{\left[\frac{\pi}{4}\left(\frac{60}{12}\right)^2\right]^2}\frac{1}{2(32.2)}Q^2 = 398 \text{ ft}$$

Solving for Q gives 143 ft³/s or 92.2 mgd.

e. c. From the solution to part **d**, one knows for the parallel pipe sections that

$$h_L = KQ^2 = K_3 Q_3^2$$

so that

$$Q_3 = Q\left(\frac{K}{K_3}\right)^{1/2} = 92.2\left(\frac{0.000594}{0.00137}\right)^{1/2} = 60.7 \text{ mgd}$$

f. b. First, the discharge in aqueduct 3 must be known in ft³/sec. By simple proportion,

$$Q_3 = 60.7\left(\frac{143}{92.2}\right) = 94.1 \text{ ft}^3/\text{sec}$$

The head loss is

$$h_L = K_3 Q_3^2 = (0.00137)(94.1)^2 = 12.1 \text{ ft}$$

g. a. The discharge in aqueduct 3 and the cross connection is the same, so

$$Q_3 = 94.1 = A_c V_c = \frac{\pi}{4}(3)^2 V_c$$

$$V_c = 13.3 \text{ ft/sec}$$

In the cross connection, the head loss is

$$h_L = f\frac{L}{D}\frac{V_c^2}{2g} = 0.012\left(\frac{20}{3}\right)\frac{(13.3)^2}{2(32.2)} = 0.22 \text{ ft}$$

This is indeed a small loss, as was assumed earlier in this set of problems.

h. b. The head loss is the difference in the energy line elevations at points D and E. The energy line elevations are

$$EL_E = 420.0 + 3.0\left(\frac{144}{62.4}\right) + \frac{V_3^2}{2g} = 427 \text{ ft}$$

$$EL_D = 46.7 + 170.0\left(\frac{144}{62.4}\right) + \frac{V_3^2}{2g} = 439 \text{ ft}$$

Thus,

$$h_L = EL_D - EL_E = 12 \text{ ft} = f\frac{L}{D}\frac{V_3^2}{2g} = f\frac{10(5280)}{78/12}\frac{(2.80)^2}{2(32.2)}$$

$$f = 0.0123$$

i. **b.** The Hazen-Williams formula is an alternative to the Darcy head loss formula; it is

$$V = 1.318 C_{HW} R^{0.63} S^{0.54}$$

The slope is $S = h_L/L$. The energy line elevations at A and B are

$$EL_A = 215.0 + 207.0\left(\frac{144}{62.4}\right) + \frac{V_2^2}{2g} = 693 \text{ ft}$$

$$EL_B = 49.0 + 179.0\left(\frac{144}{62.4}\right) + \frac{V_2^2}{2g} = 462 \text{ ft}$$

$$S = h_L/L = (EL_A - EL_B)/L = (693 - 462)/(113,000) = 0.00204$$

The hydraulic radius for a circular pipe flowing full is $R = D/4$. Substituting all data into the Hazen-Williams formula gives

$$V_2 = 7.09 = 1.318 C_{HW}\left(\frac{5}{4}\right)^{0.63}(0.00204)^{0.54}$$

with the result $C_{HW} = 133$.

j. **d.** The head loss in the cross connection is

$$h_{L_c} = EL_C - EL_D$$

The value EL_C can be obtained by extrapolating the data from points A and B to get

$$EL_C = 462 - \frac{5700}{113,000}(693 - 462) = 450.3 \text{ ft}$$

$$h_{LC} = 450.3 - 439 = 11.3 \text{ ft}$$

REFERENCES

1. Chow, V. T. *Open Channel Hydraulics.* McGraw-Hill, New York, 1959.
2. Larock, B. E., Jeppson, R. W., and Watters, G. Z. *Hydraulics of Pipeline Systems.* CRC Press, New York, 2000.
3. Munson, B. R., Young, D. F., and Okiishi, T. H. *Fundamentals of Fluid Mechanics,* 3rd ed. Wiley, New York, 1998.
4. Sanks, R. L. *Pumping Station Design.* Butterworths, Boston, 1989.
5. Street, R. L., Watters, G.Z., and Vennard, J. K. *Elementary Fluid Mechanics,* 7th ed. Wiley, New York, 1996.
6. Sturm, T. W. *Open Channel Hydraulics.* McGraw-Hill, New York, 2001.
7. White, F. M. *Fluid Mechanics,* 5th ed. McGraw-Hill, New York, 2003.

CHAPTER 6

Engineering Hydrology

PROBLEMS

6.1 A lake of modest size is the subject of study as the start of a water supply design project. The surface area of the lake is approximately 380 acres (640 acres = 1 mi^2). The data from a part of this study are presented in Exhibit 6.1. Tabulated are the monthly precipitation P, the monthly average inflow and outflow rates to and from the lake, and the water surface elevation (WSE) at the beginning of each month. The local geology suggests that subsurface seepage from the lake may be neglected.

Exhibit 6.1

Month	P, in.	$\bar{I}$, ft^3/sec	$\bar{Q}$, ft^3/sec	WSE, ft
				87.53
Feb.	3.2	36.0	29.2	
				88.21
Mar.	2.0	41.4	35.8	
				88.50
Apr.	1.2	26.3	23.4	
				87.84

a. The largest of the monthly evaporation amounts from this lake is most nearly
   a. 220 acre-feet
   b. 300 acre-feet
   c. 450 acre-feet
   d. 550 acre-feet

b. An alternative to the collection of these data is to base the determination of monthly lake evaporation on U.S. Weather Bureau pan evaporation data. Which of the following statements about the pan evaporation procedure is least correct?
   a. The value of the pan coefficient is 0.70.
   b. The value of the pan coefficient is dependent on location.
   c. The value of the pan coefficient is seasonally dependent.
   d. Pan evaporation is always larger than lake evaporation.

6.2 A rainfall hyetograph is presented in Exhibit 6.2.

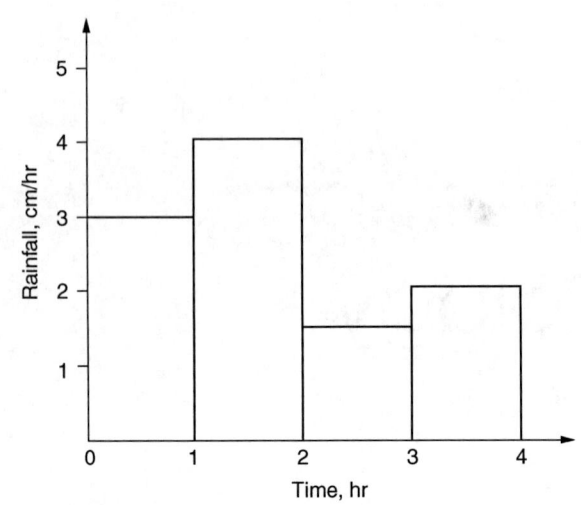

Exhibit 6.2

a. The total storm precipitation is most nearly
   a. 2.0 cm
   b. 2.6 cm
   c. 4.1 cm
   d. 10.5 cm

b. If Horton's infiltration equation with f_o = 2.0 cm/hr, f_c = 0.8 cm/hr, and k = 0.5/hr were applied to the storm described in Exhibit 6.2, the net runoff would be most nearly
   a. 2.0 cm
   b. 2.6 cm
   c. 4.1 cm
   d. 10.5 cm

c. If the net runoff from the storm were 4.0 cm, the ϕ-index that correctly describes the infiltrated amount would be nearest to
   a. 1.0 cm
   b. 1.5 cm
   c. 1.7 cm
   d. 2.3 cm

d. If the ϕ-index were actually 1.2 cm/hr, the net storm runoff in the first 2 hours would be most nearly
   a. 7.0 cm
   b. 5.6 cm
   c. 4.6 cm
   d. 3.0 cm

6.3 Exhibit 6.3 is a hydrograph of runoff from a watershed of 32 square miles as a result of a storm having an effective duration of 6 hours.

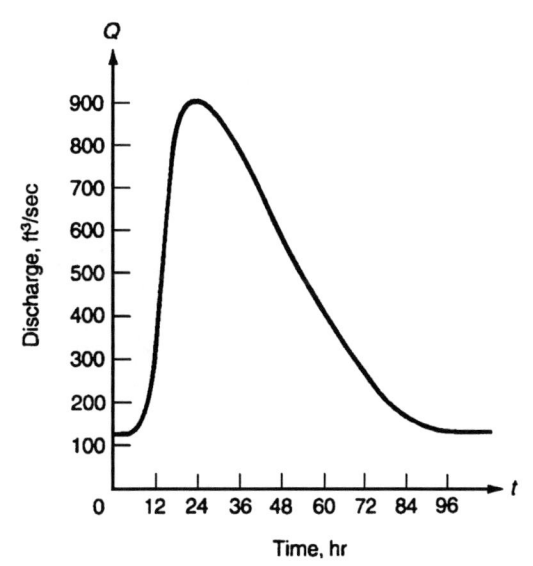

Exhibit 6.3

a. Select an appropriate method for the separation of the base flow, and develop a 6-hour unit hydrograph from the given storm hydrograph. The maximum ordinate (discharge) in the unit hydrograph is most nearly
   a. 200 ft³/sec
   b. 400 ft³/sec
   c. 600 ft³/sec
   d. 800 ft³/sec

b. Now assume that the maximum ordinate in the unit hydrograph (and other ordinates are similarly proportioned or scaled from the unit hydrograph) that was constructed in part **a** is actually 610 ft³/sec. Use this unit hydrograph to construct the storm hydrograph for a storm that produces 0.6 in. and 2.0 in. of runoff due to precipitation over two successive 6-hour periods. The maximum resulting storm discharge is most nearly
   a. 1600 ft³/sec
   b. 1550 ft³/sec
   c. 1200 ft³/sec
   d. 400 ft³/sec

6.4 The unit hydrograph for a storm of 6 hours effective duration is given in Exhibit 6.4 for a watershed having an area of 1.73 square miles.

Use the unit hydrograph to construct the storm hydrograph associated with a total precipitation of 1.0 inches in the first 6 hours and 1.5 inches in the second 6 hours. Basin losses are 0.1 inch per hour for the first 6 hours and are negligible thereafter. Assume a constant base flow of 40 ft³/sec.

The maximum storm discharge in ft³/sec and its time of occurrence after the beginning of the storm are most nearly
   a. 190 ft³/sec and 24 hours
   b. 190 ft³/sec and 18 hours
   c. 135 ft³/sec and 24 hours
   d. 90 ft³/sec and 18 hours

106 Chapter 6 Engineering Hydrology

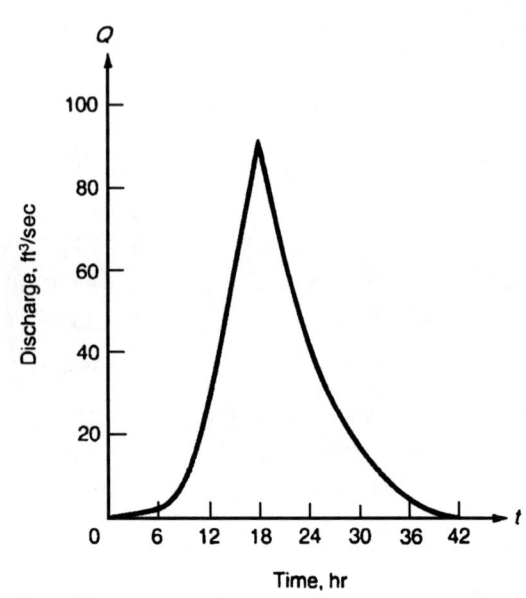

Exhibit 6.4

6.5 A detention storage pond north of town has a flat base at elevation 50.0 m. Additional elevation versus surface area data are as follows:

Elevation, m	50.0	51.0	52.0	53.0
Area, ha	0.8	2.6	3.8	4.2

The pond becomes full when the water surface elevation reaches 53.0 m.

a. The storage capacity of the detention pond is most nearly
  a. 75,000 m^3
  b. 88,000 m^3
  c. 102,000 m^3
  d. 125,000 m^3

b. At 6 AM storm runoff, described by the hydrograph in Exhibit 6.5, begins to flow into the initially empty detention pond. As a consequence of this inflow, the pond
  a. fills before 8 AM
  b. fills by approximately 8:45 AM
  c. fills by approximately 9:30 AM
  d. does not fill

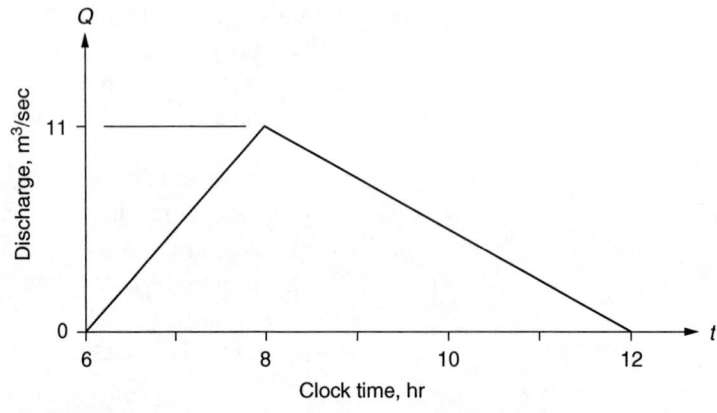

Exhibit 6.5

6.6 You are told by the project engineer on a highway construction project that, through some oversight, the area shown in Exhibit 6.6 has not been provided with a cross-highway drainage facility. The design of other pipe culverts in the area has been based on a storm that produced 0.5 in/hr of runoff during the first 30 minutes, 2.0 in/hr for the next 30 minutes, and 1.0 in/hr for the last 30 minutes. The surface of the area is mostly decomposed granite, and the velocity of flow over similar areas is about 3 ft/sec. An assortment of concrete pipes is available near the project site.

Exhibit 6.6

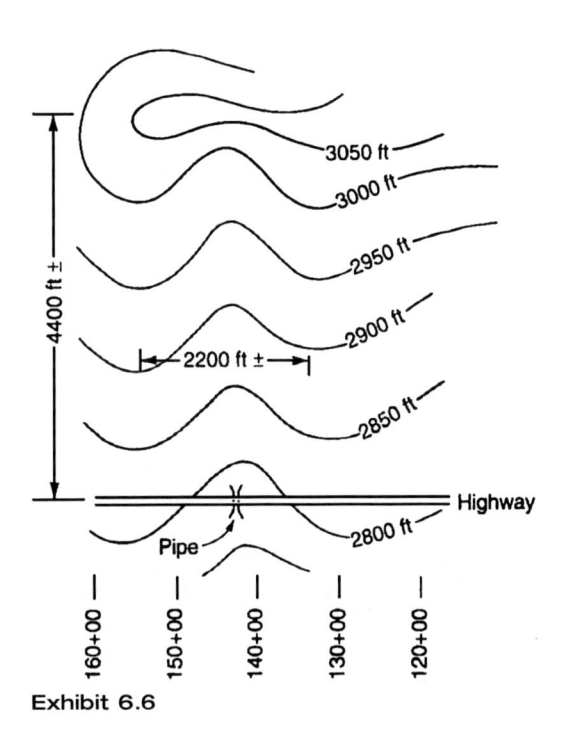

a. The peak discharge at the culvert for these conditions can be expected to be approximately
   a. 130 ft³/sec
   b. 175 ft³/sec
   c. 220 ft³/sec
   d. 265 ft³/sec

b. A highway cross section (viewed from the right, with flow from right to left) is shown in Exhibit 6.6(a). Assume a free outlet and a water surface at the inlet that is the same elevation as the top of the pipe. The most appropriate size of concrete pipe for this inlet would be
   a. 60 in
   b. 54 in
   c. 48 in
   d. 42 in

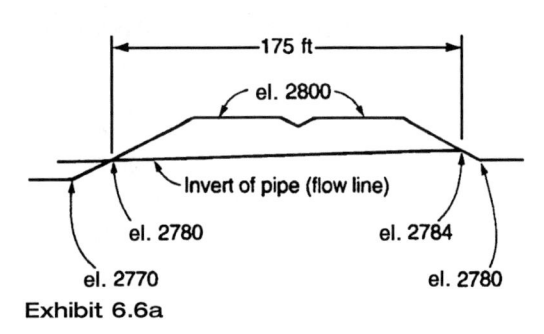

Exhibit 6.6a

6.7 The inflow and outflow hydrographs for a river reach are tabulated below.

Day	Time	Inflow, m³/sec	Outflow, m³/sec
1	12 AM	160.0	57.0
	12 PM	160.0	57.0
2	12 AM	375.0	170.0
	12 PM	225.0	240.0
3	12 AM	150.0	220.0
	12 PM	110.0	180.0
4	12 AM	85.0	140.0
	12 PM	62.0	105.0

The following is an additional inflow hydrograph for this river reach:

Day	Time	Inflow, m³/sec
1	12 AM	20.0
	12 PM	200.0
2	12 AM	425.0
	12 PM	255.0
3	12 AM	170.0
	12 PM	140.0

a. First, determine the storage in 12-hour intervals. Then use these data to select the Muskingum routing coefficients x and K. The most appropriate pair of routing coefficients for this river reach are
  a. $x = 0.1$, $K = 0.79$ day
  b. $x = 0.2$, $K = 0.70$ day
  c. $x = 0.1$, $K = 1.27$ days
  d. $x = 0.2$, $K = 1.43$ days

b. Use the Muskingum routing method, a 12-hour time interval, and the (x, K) pair chosen in part **a** to determine the magnitude Q_m of the maximum outflow from the river reach for the given inflow sequence, assuming uniform flow at 12 AM on the first day. The value of Q_m is most nearly
  a. 180 m³/sec
  b. 210 m³/sec
  c. 240 m³/sec
  d. 270 m³/sec

6.8 Data from a well test have been collected and plotted on semilogarithmic paper. A fit to the straight-line portion of the data passes through the points ($t = 10^{-3}$ day, $s = 3.5$ ft) and ($t = 10^{-2}$ day, $s = 9.0$ ft), in which t is time and s is drawdown. The data were collected at an observation well located 100 feet from a well that is pumped at 1.0 ft³/sec in a confined aquifer.

a. For this aquifer, the transmissibility T is most nearly
  a. 21,500 gallons/day/ft
  b. 2900 gallons/day/ft
  c. 900 gallons/day/ft
  d. 120 gallons/day/ft

b. The storage constant for the aquifer is most nearly
  a. 6.6×10^{-5}
  b. 1.5×10^{-4}
  c. 6.0×10^{-4}
  d. 1.1×10^{-3}

c. The drawdown 100 feet from the pumped well after 24 hours of pumping will be most nearly
  a. 25.0 ft
  b. 23.0 ft
  c. 20.0 ft
  d. 17.0 ft

6.9 A confined aquifer with a transmissibility $T = 40$ m²/day lies under rangeland with an average ground surface elevation of 150 m. A well was installed long ago in this aquifer; currently it is operated continually at a steady rate. Two observation wells located 300 m and 600 m from the pumped well have water surface elevations of 138.4 m and 141.8 m, respectively.

a. The discharge is most nearly
  a. 4.5 liters/sec
  b. 7.1 liters/sec
  c. 14.3 liters/sec
  d. 32.4 liters/sec

b. Assuming that the pump has an efficiency of 80 percent and that the well diameter is 30 cm, the power output of the pump to bring the water to the surface is most nearly
  a. 6.9 kW
  b. 8.6 kW
  c. 9.3 kW
  d. 11.6 kW

SOLUTIONS

6.1

a. c. Exhibit 6.1a is a schematic diagram of the lake, showing lake surface area A, inflow I, outflow Q, precipitation P, and the neglect of any flow to groundwater $G \approx 0$. A monthly volume balance approach is applied to the lake in the form

$$PA + (\bar{I} - \bar{Q})\Delta t - E = \Delta S$$

with the common units for each term to be acre-ft. Hence, some additional unit conversion factors are needed:

$$PA = \left(\frac{P(\text{in})}{12}\right)A$$

$$(\bar{I} - \bar{Q})\Delta t = (\bar{I} - \bar{Q})\frac{\text{ft}^3}{\text{sec}} \frac{(60)^2(24)N \text{ sec/month}}{43,560 \text{ ft}^3/\text{acre-ft}}$$

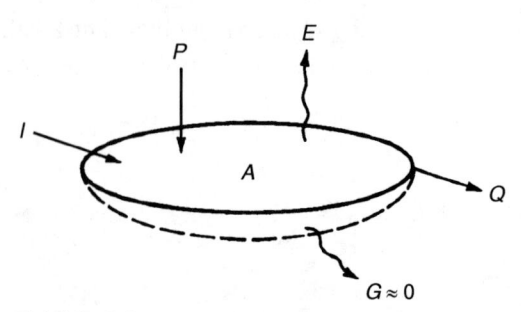

in which N is the number of days in a particular month,

$$\Delta S = A \times \Delta \text{WSE}$$

with A in acres. The computations can be summarized in the following table:

Month	N, days	PA, acre-ft	$(\bar{I} + \bar{Q})\Delta t$, acre-ft	ΔS, acre-ft	E, acre-ft
Feb.	28	101	378	258	221
Mar.	31	63	344	110	297
Apr.	30	38	173	−251	462

The monthly evaporation increases as the months become warmer.

b. a. The U.S. Weather Bureau Class A pan is a 4-foot-diameter metal tank in which water is kept at a depth of 8 inches. Lake evaporation E is determined from pan evaporation E_p by the relation $E = c_p E_p$.

If the pan coefficient c_p is in fact a known constant for this particular application, then the use of pan data to determine lake evaporation is definitely the easier procedure. However, in practice, this coefficient is often chosen to be 0.7 simply because this is the approximate average value to emerge from other studies and applications. In addition, the value of this coefficient is known to be location-dependent and also to change with the season of the year. The primary reason for this behavior is clear enough: a small tank cannot fully mimic the thermodynamic behavior of the much larger body of water. If pan data were collected daily over the same period as the other study data, it would be possible to determine the appropriate monthly pan coefficient values for use at this site in future years.

6.2

a. d. The total precipitation is the area under the rainfall versus time plot. Since the data are given in hourly increments, the total is a simple sum $P = 3.0 + 4.0 + 1.5 + 2.0 = 10.5$ cm.

b. b. If the given parameters are inserted into Horton's expression [Eq. (6.2)] for infiltration f, one has $f = 0.8 + 1.2e^{-0.5t}$. The infiltrated amount is then

$$V = \int_0^T f\,dt = \int_0^T [f 0.8 + 1.2 e^{-0.5t}]\,dt = 0.8T + \frac{1.2}{0.5}[1 - e^{-0.5T}]$$

$$= 0.8(4.0) + \frac{1.2}{0.5}[1 - e^{-0.5(4.0)}] = 5.28 \text{ cm}$$

The runoff R is the difference between the total precipitation and the infiltrated volume, or $R = 10.5 - 5.28 = 5.22$ cm.

c. **c.** The runoff is the portion of the hyetograph that lies above the ϕ-index. Begin by estimating ϕ, say $\phi < 1.5$ cm. Then,

$$4.0 = \sum_i [p_i - \phi] = (3.0 - \phi) + (4.0 - \phi) + (1.5 - \phi) + (2.0 - \phi)$$

or $4.0 = 10.5 - 4\phi$ and $\phi = 1.625$, which is contrary to the starting assumption for ϕ. Consequently, assume 1.5 cm $< \phi < 2.0$ cm, leading to

$$4.0 = \sum_i [p_i - \phi] = (3.0 - \phi) + (4.0 - \phi) + (2.0 - \phi)$$

Thus, $4.0 = 9.0 - 3\phi$, and $\phi = 1.67$, which is in agreement with the estimate.

d. **c.** In this case,

$$R = (3.0 - \phi) + (4.0 - \phi) = 7.0 - 2(1.2) = 4.6 \text{ cm}$$

6.3

a. **c.** The first step is to select a way to separate the base flow, which is streamflow not caused directly by the storm, from the rest of the flow. This task can be done in several ways, but this time, the method that uses the equation $N = A^{0.2}$ (A in square miles) will be chosen. The duration of the storm recession curve is therefore

$$N = A^{0.2} = (32)^{0.2} = 2.0 \text{ days}$$

This base flow separation is shown in Exhibit 6.3a. Although the choice of a constant base flow of 125 ft^3/sec is a possible alternative, it would have the effect of extending the duration of the unitgraph by another 24 hours.

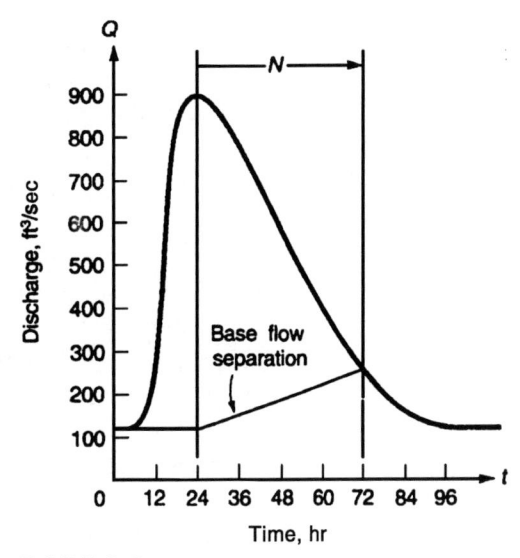

Exhibit 6.3a

Time, hr	Base Flow, ft³/s	Q_{total}, ft³/s	Q_{storm}, ft³/s	UH Ordinates, ft³/s
(1)	(2)	(3)	(4)	(5)
0	125	125	0	0
6	125	125	0	0
12	125	250	125	98
18	125	800	675	531
24	125	900	775	610
30	142	850	708	557
36	157	765	608	479
42	174	675	501	394
48	190	585	395	311
54	206	500	294	321
60	222	405	183	144
66	239	335	96	76
72	255	255	0	0
			$\Sigma = 4360$	

Columns 1–3 are data taken from Exhibit 6.3a. The storm, and the outflow from it, begin at a time of 6 hours. (The base flow line that is projected to the point under the peak flow cannot be drawn well if some prior base flow history is not given.) Column 4 is column 3 minus column 2.

Using the sum of ordinates of column 4 in Eq. (6.5) in *Civil Engineering PE License Review*, the storm runoff volume is $V = 4360$ ft³/sec-6 hrs. The equivalent water depth over the watershed is

$$x = \frac{V}{A} = \frac{(4360 \text{ ft}^3/\text{sec})(6 \text{ hr})(60)^2 \text{ sec/hr}}{(32 \text{ mi}^2)(5280 \text{ ft/mi})^2} = 0.106 \text{ ft} = 1.27 \text{ in}$$

The unit hydrograph ordinates listed in column 5 are found by dividing the ordinates in column 4 by 1.27. The unit hydrograph is plotted in Exhibit 6.3b with a time shift of 6 hours, so a time of zero is at the beginning of the plot.

Although the question only asks for the maximum unit hydrograph ordinate, one must resist the temptation to try to determine this value without developing the entire unit hydrograph; there is no way to do this.

b. **b.** The ordinates of the composite storm hydrograph are most easily developed in a table, as shown.

Observe that the second part of the composite storm hydrograph, column 4, begins 6 hours after the first part, column 3, and the composite storm ordinates Q are simply the sum of the two previous columns. The composite storm hydrograph is plotted in Exhibit 6.3c. For completeness, in an actual application, a suitable base flow must be added to these values.

Although Exhibits 6.2b and 6.2c are included for completeness, the problems can be completed without actually drawing either of them, since the desired information may be taken directly from the tabular computations.

Time, hr	UH Ordinates, ft³/sec	0.6 × UH, ft³/sec	2.0 × UH, ft³/sec	Q, ft³/sec
0	0	0	—	0
6	98	59	0	59
12	531	319	196	515
18	610	366	1062	1428
24	557	334	1220	1554
30	479	287	1114	1401
36	394	236	958	1194
42	311	187	788	975
48	231	139	622	761
54	144	86	462	548
60	76	46	288	334
66	0	0	152	152
72			0	0

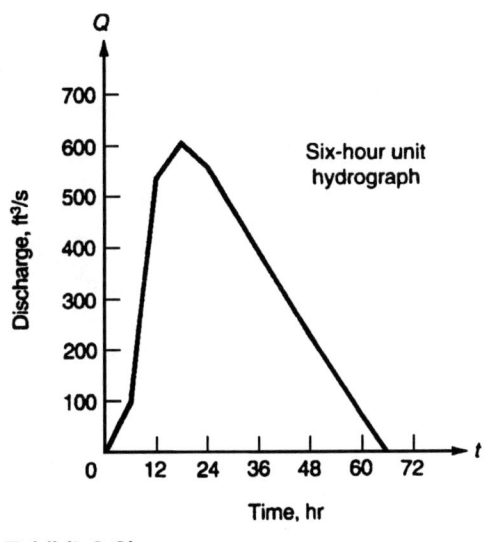

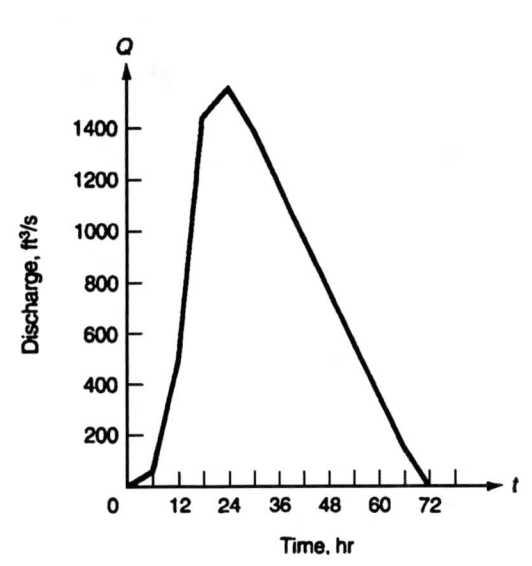

Exhibit 6.3b

Exhibit 6.3c

6.4

a. In the first 6 hours, the net precipitation is $P_1 = 1.0 - 0.1(6) = 0.4$ in. In the second 6 hours, the net precipitation is $P_2 = 1.5$ in. The computation of ordinates for the storm hydrograph is presented in the table that follows; an explanation of the table entries follows the table.

The values in column 2 are unit hydrograph discharges read from the graph in the problem statement at the times listed in column 1. Columns 3 and 4 are the contributions to the storm hydrograph from the net precipitation in the first and second 6-hour intervals. The base flow is listed in column 5. The storm hydrograph in column 6 is the sum of the component values in columns 3–5.

Time, hr	U_i, ft³/sec	P_1U_i, ft³/sec	P_2U_i, Lagged, ft³/sec	Base Flow, ft³/sec	Q_{storm} ft³/sec
(1)	(2)	(3)	(4)	(5)	(6)
0	0	0	—	40	40
6	2	1	0	40	41
12	30	12	3	40	55
18	90	36	45	40	121
24	41	16	135	40	191
30	17	7	62	40	109
36	6	2	26	40	68
42	0	0	9	40	49
48			0	40	40

6.5

a. b. The storage computations can be summarized in the following table:

Elevation, m	Area, ha	ΔS, ha-m	S, ha-m
50.0	0.8		0.0
		1.7	
51.0	2.6		1.7
		3.2	
52.0	3.8		4.9
		3.9	
53.0	4.0		8.8

The first two columns repeat the known data; observe that the increment in elevation is 1.0 m. The increment in storage, ΔS, listed in the third column between two elevations, is then

$$\Delta S = \frac{1}{2}(A_i + A_{i+1})(1.0)$$

and the storage itself is the sum of the incremental storage values from the base value of zero up to the elevation of interest. In this case, the pond capacity is 8.8 ha-m = 88,000 m³.

b. c. The storm inflow volume can be computed directly from Exhibit 6.5 as

$$V = \frac{1}{2}\left(11\frac{m^3}{sec}\right)(6\ hr) = 33(3600)\ m^3/sec = 118{,}800\ m^3$$

Since this volume considerably exceeds the storage capacity of the pond, the pond will fill.

With the aid of Exhibit 6.5a, one can calculate the time when filling is completed. The pond volume filled by 8 AM is $V_1 = 2(11.0/2) = 11.0$ (m³/s)-hr = 11.0(3600) = 39,600 m³. The volume that remains to be filled is then $V_2 = 88{,}000 - 39{,}600 = 48{,}400$ m³. Letting t be the clock

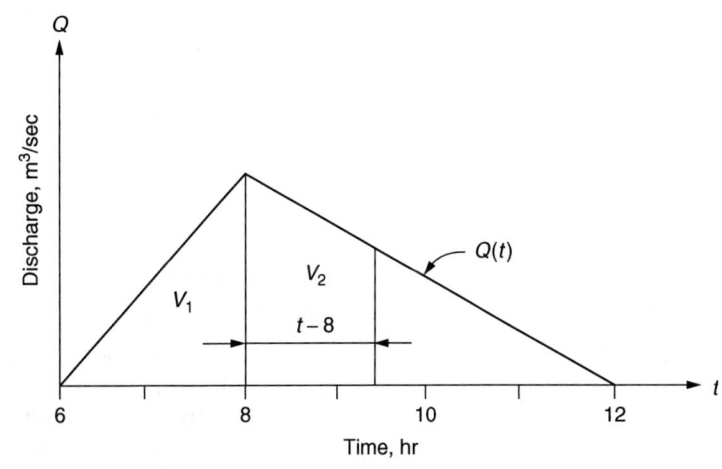

Exhibit 6.5a

time, the discharge $Q(t)$ in Exhibit 6.5a may be written as $Q(t) = 11.0[12 - t]/4$ for $8 \leq t \leq 12$. Then V_2 may be written as

$$V_2 = \frac{1}{2}[Q(8) + Q(t)](t - 8)$$

or, upon substitution for $Q(t)$ and some rearrangement, $9.78 = (16 - t)(t - 8)$. In standard form, $t^2 - 24t + 137.8 = 0$. Hence,

$$t = \frac{1}{2}\{24 - [24^2 - 4(137.8)]^{1/2}\} = 9.51 \text{ hr} \approx 9:30$$

The detention pond is filled at 9:30 AM.

At the sacrifice of some accuracy but possibly with some gain in solution speed, this question could be approached in a way that is much like the tabular solution of the first question in this problem. The computational table could be set up as shown in Exhibit 6.5b.

The third column lists the average discharge for the time interval. The incremental volume, column 4, that is filled during that time interval is $\Delta V = \overline{Q}\Delta t = 3600\overline{Q}$ m³ for a time interval of 1 hour. The volume that has

Exhibit 6.5b

Time i, hr	Q, m³/s	$\overline{Q}$, m³/s	ΔV, m³	$V = \Sigma \Delta V$, m³
6	0			0
		2.75	9,900	
7	5.50			9,900
		8.25	29,700	
8	11.00			39,600
		9.63	34,650	
9	8.25			74,250
		6.88	24,770	
10	5.50			99,029
11				
12				

flowed into the pond is accumulated in the last column until that volume exceeds the capacity of the pond. Then there is no need to complete the remainder of the table. One can now conclude that the pond is filled after 9 AM and before 10 AM.

6.6

a. c. In this problem, sufficient information is supplied to allow one to apply the rational method to estimate the peak discharge. It is well known that this method becomes more questionable as the area to which it is applied increases. However, the result is usually acceptable if the area is less than 1 square mile, which is the case here.

First, the time of concentration t_c for the tributary area must be estimated. Inspecting Exhibit 6.6, the contours are such that the longest flow path will be from the upper left or right corner of the map to the culvert, roughly a distance $L = (4400^2 + 1100^2)^{1/2} = 4540$ ft, which leads to a time of concentration

$$t_c = \frac{L}{V} = \frac{4540}{3(60)} = 25.2 \text{ min}$$

Since this value is less than 30 minutes, the full area will contribute to the peak discharge within each half-hour precipitation interval, and the appropriate intensity to choose is 2.0 in/hr. Selecting a correct runoff coefficient is somewhat of a problem, since Table 6.1 in *Civil Engineering PE License Review* does not list decomposed granite itself. Since this material is expected to be more porous than pavement and less porous than lawn (!), an intermediate value of 0.5 will be chosen here. Thus, the rational formula gives the peak discharge as

$$Q_p = CiA = (0.5)(2.0)\frac{(2200)(4400)}{43,560} = 222 \text{ acre-in/hr} \approx 222 \text{ ft}^3/\text{sec}$$

b. b. One might begin the selection process by temporarily assuming that the pipe flows full and applying the Manning equation

$$Q = \frac{1.49}{n} AR^{2/3} S^{1/2}$$

From Exhibit 6.6(a), the slope is $S = 4/175 = 0.0229$. The area A and hydraulic radius R are

$$A = \frac{\pi}{4}d^2 \qquad R = \frac{A}{P} = \frac{\pi d^2/4}{\pi d} = \frac{d}{4}$$

for a pipe flowing full. From Table 5.2 in *Civil Engineering PE License Review*, values of the Manning roughness factor n can range from 0.011 to 0.023, depending on the surface finish; in this case, $n = 0.015$ seems reasonable. Thus,

$$Q = 222 = \frac{1.49}{0.015}\left(\frac{\pi}{4}d^2\right)\left(\frac{d}{4}\right)^{2/3}(0.0229)^{1/2}$$

$$d^{8/3} = 47.4 \text{ ft}^{8/3}$$

$$d = (47.4)^{3/8} = 4.25 \text{ ft} = 51 \text{ in}$$

In this size range, the usual commercial pipe-size increment is 3 inches, so a 51 in. diameter would appear to be marginally acceptable on the basis of this analysis. Some contend that a 48 in. diameter would be adequate, but without more information, such as a frequency analysis and better information on the runoff coefficient, a more conservative approach seems warranted. And the pipe is available near the site!

Since the slope is rather steep, it does not appear that the culvert will actually flow full. A Bureau of Public Roads pipe flow chart indicates for 54 in. pipe and $n = 0.015$, the normal depth of flow would be about 3.1 ft with a critical slope of 0.005. Here, S exceeds the critical slope, so the flow will be supercritical through most of the culvert and a 54 in. pipe diameter will be adequate. The BPR chart further indicates that a 48 in. pipe can carry at most 200 ft^3/sec for the given conditions.

6.7

a. **a.** The computed results for storage S and weighted discharge values $W = xI + (1 - x)Q$ for $x = 0.1$ and $x = 0.2$ will be collected together in one table. The computation of storage begins with the fundamental relation, Eq. (6.7) in the text, which is

$$I - Q = \frac{dS}{dt}$$

Integrating this relation between two time instants and applying the trapezoidal rule to evaluate the integral leads to

$$\Delta S = \frac{\Delta t}{2}[I_1 + I_2 - (Q_1 + Q_2)]$$

and

$$S_{i+1} = S_i + \Delta S$$

with $\Delta t = t_2 - t_1$. For example, the storage at 12 PM on Day 1 is

$$S_2 = S_1 + \Delta S = 0 + \frac{\frac{1}{2}}{2}[30.0 + 160.0 - (30.0 + 57.0)] = 25.75 \text{ m}^3/\text{sec-day}$$

using $\Delta t = 12$ hr $= \frac{1}{2}$ day. The units are m^3/sec-day, representing 1 m^3/sec for 24 hours, a volume. The values of channel storage and of weighted discharge $W = x(I - Q) + Q$ are listed in the following table:

Day	Time	Inflow I, m^3/sec	Outflow Q, m^3/sec	Storage S, m^3/sec-days	W, $x = 0.1$, m^3/sec	W, $x = 0.2$, m^3/sec
1	12 AM	30.0	30.0	0	30.0	30.0
	12 PM	160.0	57.0	25.75	67.3	77.6
2	12 AM	375.0	170.0	102.75	190.5	211.0
	12 PM	225.0	240.0	150.25	238.5	237.0
3	12 AM	150.0	220.0	129.00	213.0	206.0
	12 PM	110.0	180.0	94.00	173.0	166.0
4	12 AM	85.0	140.0	62.75	134.5	129.0
	12 PM	62.0	105.0	38.25	100.7	96.4

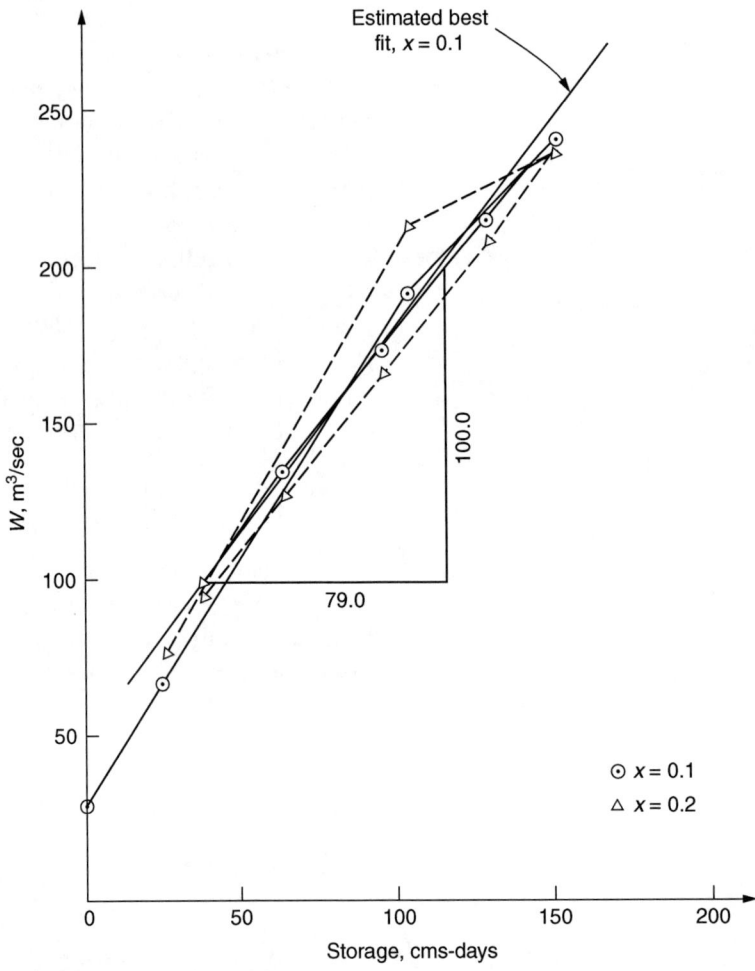

Exhibit 6.7

The results in the table were used to create Exhibit 6.7, which is a plot of weighted discharge W versus storage S for two values of x. From the plot, it seems clear that $x = 0.1$ produces a narrower envelope of data. Hence, a straight line was fitted by eye to the plot for $x = 0.1$. The parameter K was then found to be

$$\frac{1}{K} = \frac{100.0}{79.0}, \quad K = 0.79 \text{ day}$$

b. **d.** The Muskingum routing equation, Eq. (6.11) in *Civil Engineering PE License Review*, is

$$Q_{n+1} = C_0 I_{n+1} + C_1 I_n + C_2 Q_n$$

This equation is efficiently evaluated with the aid of a table, but first the coefficients in the equation must be evaluated with the data for the problem. The denominator in the coefficients is $D = K - Kx + \Delta t/2 = 0.79 - 0.79(0.1) + 0.5/2 = 0.961$, and the other coefficients are then

$$C_0 = (\Delta t/2 - Kx)/D = [0.25 - 0.79(0.1)]/0.961 = 0.178$$

$$C_1 = (\Delta t/2 + Kx)/D = [0.25 + 0.79(0.1)]/0.961 = 0.342$$

and

$$C_2 = (K - Kx - \Delta t/2)/D = [0.79 - 0.79(0.1) - 0.25]/0.961 = 0.480$$

Now a final routing table will be used to display the results from the evaluation of Eq. (6.11):

Day	Time	I_n, m³/sec	$C_0 I_{n+1}$	$C_1 I_n$	$C_2 Q_n$	Q_{n+1}, m³/sec
1	12 AM	20.0				20.0
	12 PM	200.0	35.6	6.8	9.6	52.0
2	12 AM	425.0	75.7	68.4	25.0	169.1
	12 PM	255.0	45.4	145.4	81.2	272.0
3	12 AM	170.0	30.3	87.2	130.6	248.1

The outflow at the end of the time interval, Q_{n+1}, which appears in the last column, is simply the sum of the entries in the three preceding columns. The outflow will continue to decrease with time, so there is no reason to continue the calculations. Thus, the peak outflow is 272.0 m³/s occurring near 12 PM on the second day.

6.8

a. **a.** The semilogarithmic plot was evidently prepared as a prelude to the use of the equations of the Cooper-Jacob method, which are Eqs. (6.23) and (6.24) in Chapter 6 of *Civil Engineering PE License Review*. Extrapolation of the straight line to a drawdown of zero will yield the needed value of t_0.

The straight line may be written as $y - y_1 = m(x - x_1)$, where the x values are the plotted logarithmic values. To determine the slope m, use both data pairs: $9.0 - 3.5 = m(\log_{10} 10^{-2} - \log_{10} 10^{-3})$ and $m = 5.5$. To determine t_0 when the drawdown is zero,

$$0 - 3.5 = 5.5(\log_{10} t_0 - \log_{10} 10^{-3})$$

$$\log_{10} t_0 = -3.0 - \frac{3.5}{5.5} = -3.636$$

$$t_0 = 10^{-3.636} = 2.3 \times 10^{-4} \text{ day}$$

The transmissibility computed using the Cooper-Jacob method is

$$T = \frac{2.3Q}{4\pi(s_2 - s_1)} \log_{10}\left(\frac{t_2}{t_1}\right) = \frac{2.3(1.0)}{4\pi(9.0 - 3.5)} \log_{10}\left(\frac{10^{-2}}{10^{-3}}\right)$$

$$T = 0.0333 \frac{\text{ft}^2}{\text{sec}} = \left[0.0333 \frac{\text{ft}^2}{\text{sec}}\right]\left[7.48 \frac{\text{gal}}{\text{ft}^3}\right]\left[60^2 \times 24 \frac{\text{sec}}{\text{day}}\right] = 21,500 \frac{\text{gal/day}}{\text{ft}}$$

b. **b.** The storage coefficient computed using the Cooper-Jacob method is

$$S = \frac{2.25 T t_0}{r^2} = \frac{2.25 \left(\frac{21,500}{7.48} \frac{\text{ft}^2}{\text{day}}\right)(2.3 \times 10^{-4} \text{day})}{(100 \text{ ft})^2} = 1.49 \times 10^{-4}$$

c. **c.** With the data that are now available, it is straightforward to evaluate the Theis equations directly. Equation (6.21) gives

$$u = \frac{r^2 S}{4Tt} = \frac{(100 \text{ ft})^2 (1.49 \times 10^{-4})}{4\left[\frac{21{,}500}{7.48} \frac{\text{ft}^2}{\text{day}}\right][1 \text{ day}]} = 1.3 \times 10^{-4}$$

Using linear interpolation in Table 6.2 yields $W(u) = 8.42$ and, using Eq. (6.20),

$$s = \frac{Q}{4\pi T} W(u) = \frac{1.0 \text{ ft}^3/\text{sec}}{4\pi (0.0333 \text{ ft}^2/\text{sec})} (8.42) = 20.1 \text{ ft}$$

6.9

a. **c.** For a confined aquifer

$$Q = \frac{2\pi T(h_2 - h_1)}{\ln(r_2/r_1)} = \frac{2\pi\left[\frac{40}{60^2 \times 24} \frac{\text{m}^3}{\text{s}}\right](141.8 - 138.4)}{\ln(600/300)}$$

and $Q = 0.0143 \text{ m}^3/\text{s} = 14.3$ liters/s.

b. **b.** The discharge equation for a confined aquifer can also be used to find the piezometric head at the well:

$$Q = 0.0143 = \frac{2\pi T(h_2 - h_1)}{\ln(r_2/r_1)} = \frac{2\pi\left[\frac{40}{60^2 \times 24}\right](138.4 - h_w)}{\ln(300/0.15)}$$

yielding $h_w = 100.7$ m. The pumping lift to the surface is then $150 - 100.7 = 49.3$ m. The power required to achieve this lift is

$$\frac{1}{0.80} Q\gamma h = \frac{1}{0.80}(0.0143)(9800)(49.3) = 8.64 \text{ kW}$$

REFERENCES

1. ASCE. *Design and Construction of Sanitary and Storm Sewers* (Manual No. 37). American Society of Civil Engineers, New York, 1986.
2. Bedient, P. B., and Huber, W. C. *Hydrology and Floodplain Analysis*, 3rd ed. Addison-Wesley, Reading, MA, 2002.
3. Chow, V. T., Maidment, D. R., and Mays, L. R. *Applied Hydrology*. McGraw-Hill, New York, 1988.
4. Linsley, R. K., Kohler, M. A., and Paulhus, J. L. H. *Hydrology for Engineers*, 2nd ed. McGraw-Hill, New York, 1975.
5. McCuen, R. H. *Hydrologic Analysis and Design*, 2nd ed. Prentice-Hall, Englewood Cliffs, NJ, 1998.
6. Viessman, Jr., W., and Lewis, G. L. *Introduction to Hydrology*, 4th ed. Addison-Wesley, Reading, MA, 1996.

CHAPTER 7

Water Quality, Treatment, and Distribution

PROBLEMS

7.1 A city uses a nearby river as a water source and pumps the water to the distribution system and to a storage tank. The required hourly flow rates for the maximum daily use are

Time of Day	Flow Rate (gpm)
12 PM	0
2 AM	4000
4 AM	8000
6 AM	12,000
8 AM	12,000
10 AM	12,000
12 noon	12,000
2 PM	12,000
4 PM	12,000
6 PM	12,000
8 PM	8000
10 PM	4000

    a. The maximum daily flow rate required for this system is
        a. 6000 gpm
        b. 8000 gpm
        c. 9000 gpm
        d. 10,000 gpm

b. The storage requirements to supply the maximum daily flow, if a constant pumping rate is set at 10,000 gpm, are
   a. more than 6 MG
   b. from 4 to 6 MG
   c. from 2 to 4 MG
   d. from 1 to 2 MG

c. If the fire flow is determined to be 6000 gpm, then
   a. the design flow would be 15,000 gpm
   b. the design flow could be less than, equal to, or greater than 15,000 gpm
   c. the required duration for the fire flow would be about 6 hours
   d. both (a) and (c)

d. If a pumping rate of 8000 gpm was used on the maximum day, the storage tank would fill
   a. from 6 AM until 8 PM
   b. from 4 AM until 8 PM
   c. from 8 PM until 6 AM
   d. from 8 PM until 4 AM

e. If the elevation of the load center is 40 ft above the water source, the height of the water above the water source in the storage system should be
   a. from 40 to 60 ft
   b. from 60 to 80 ft
   c. from 80 to 100 ft
   d. more than 100 ft

f. This water distribution system would serve about how many persons?
   a. 5000
   b. 10,000
   c. 50,000
   d. 100,000

g. If the capacity of each pump at the water treatment plant is 2500 gpm, the number of pumps required is
   a. 4 plus 1 spare
   b. 4 plus 2 spares
   c. 5 plus 1 spare
   d. 5 plus 2 spares

h. If the capacity of each pump at the water treatment plant is 2500 gpm and the load center is 40 ft above the water source, the horsepower requirement for each pump is about
   a. 30 hp
   b. 60 hp
   c. 100 hp
   d. 150 hp

i. Fire-flow rates would be determined based upon
   a. climate
   b. type of construction
   c. floor area of buildings
   d. (b) and (c)

j. The minimum pressure at the load center would occur
   a. under maximum daily flow
   b. under peak hourly flow
   c. under the sum of the maximum daily flow and the fire flow
   d. under (b) or (c)

7.2 A major flood has resulted in contamination of hundreds of domestic water wells and the distribution system of a small town.

a. The principal contaminants of concern that affect the use of a well and the distribution system are
   a. viruses
   b. bacteria
   c. suspended solids
   d. all of the above

b. The chlorine dose required to disinfect a contaminated well would be approximately
   a. 1 mg/L
   b. 5 mg/L
   c. 10 mg/L
   d. 50 mg/L

c. If a 5% sodium hypochlorite solution is used for disinfection, the quantity needed to disinfect 200 ft of 2 in diameter pipe in the well is
   a. 0.1 gal
   b. 0.15 gal
   c. 0.3 gal
   d. 0.5 gal

d. If the bacterial decay constant is 0.1/min, the time necessary to get a kill of 1 in 1×10^6 is about
   a. 15 min
   b. 30 min
   c. 60 min
   d. 120 min

e. The pH should be controlled to less than 7 because
   a. it reduces the growth of viruses
   b. a pH of less than 7 will kill bacteria
   c. the chlorine will be in the HOCl form primarily
   d. the chlorine will be in the Cl_2 form primarily

f. Doubling the chlorine dosage will
   a. result in a much lower required contact time
   b. result in one-half the required contact time
   c. result in twice the required contact time
   d. have an unknown effect

g. Important inorganic constituents that could affect the effectiveness of chlorine would include
   a. carbonate
   b. iron
   c. manganese
   d. all of the above

h. The most desirable chlorine compound to use in this situation would be
   a. 5% chlorine bleach
   b. 10% chlorine bleach
   c. 70% high-test hypochlorite
   d. chlorine gas

i. Procedures for collecting water samples to determine bacteriological quality include
   a. using a sterile bottle
   b. delivering the samples to the lab within one day
   c. using a cool bottle during transport
   d. all of the above

j. The addition of sodium hypochlorite would
   a. raise the pH
   b. lower the pH
   c. have no effect on the pH
   d. form Cl_2 primarily

7.3 A city with an expected increase in its future population is to provide a complete water treatment plant for a surface water source, including rapid mix, flocculation, sedimentation, filtration, and chlorination. The design flow is 22.5 MGD.

a. Assuming alum is used for coagulation, the approximate area of the sedimentation basin
   a. is closest to 10,000 ft^2
   b. is closest to 20,000 ft^2
   c. depends on the dosage of alum
   d. depends on the quality of the water source

b. The basin for the rapid mixing of the alum would
   a. require a detention time of about 1 min
   b. require a detention time of about 60 min
   c. require a size dependent upon the alum dosage
   d. strongly affect the success of the flocculation step

c. The dosage of alum would probably
   a. be greater than 100 mg/L
   b. lower the pH of the water
   c. result in an aluminum sulfate sludge
   d. raise the pH of the water

d. The required detention time of the flocculation unit would be approximately
   a. 15 min
   b. 20 min
   c. 60 min
   d. Depends on the dosage of alum

e. The required area of the sand filter assuming a single sand media
   a. would be closest to 200 ft^2
   b. would be closest to 500 ft^2
   c. depends on water quality
   d. depends on alum dosage

f. If THM precursors were not a concern for this plant, disinfectants would probably be added
   a. at a rapid mix basin
   b. at a flocculation basin
   c. after the sand filtration
   d. (a) and (c)

g. Waste sludges from this plant would be composed of
   a. calcium hydroxide
   b. aluminum sulfate
   c. dissolved solids from the water source
   d. suspended solids from the water source

h. Assuming adequate storage in the distribution system, the maximum daily flow would be used to design
   a. the sand filter
   b. the coagulation basin
   c. the rapid mix basin
   d. none of these

i. Elevated storage at the plant could be required for
   a. the chlorinators
   b. the backwash sand filters
   c. firefighting
   d. grounds watering

j. Control of taste and odor problems
   a. may require a strong oxidant
   b. may require activated carbon
   c. will require a separate unit process
   d. (a) and (b)

SOLUTIONS

7.1 **a. c.** The maximum daily flow is determined from integration of the flow rate versus time data given, as shown in Exhibit 7.1, for the day of maximum use.

$$Q^{daily} = \frac{216{,}000 \text{ gpm-h}}{24 \text{ h}}$$

$$= 9000 \text{ gpm}$$

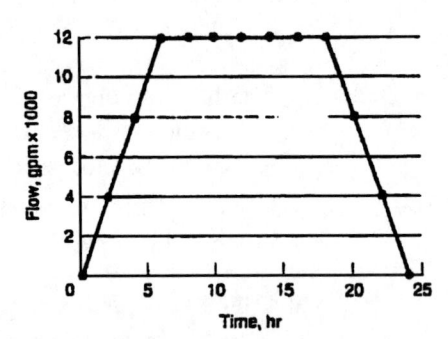

Exhibit 7.1

b. d. If pumping can supply 10,000 gpm, then the total storage requirement is shown in the hatched area of Exhibit 7.1(a).

$$S = (2000 \text{ gpm-h}(13 \text{ hr}))\left(\frac{60 \text{ min}}{1 \text{ h}}\right)$$

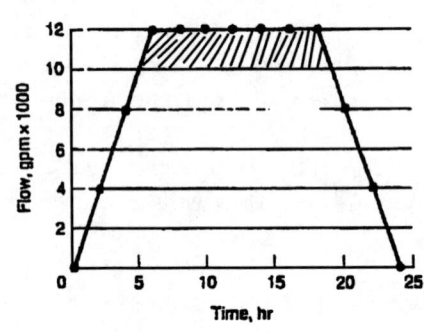

Exhibit 7.1(a)

c. d. Since the maximum hourly flow is less than 15,000 gpm, the design flow is the sum of the maximum daily flow and the fire flow. The flow is 15,000 gpm, and the duration required for the fire flow would be 6 hr.

d. d.

e. d. The pressure in the distribution system should be about 40 psi under static conditions. As such, the storage elevation should be

$$H_s = 40 \text{ ft} + (40 \text{ psig})\left(\frac{1 \text{ ft}}{0.43 \text{ psig}}\right)$$

$$= 133 \text{ ft}$$

f. **b.** Using a per capita flow of 100 gal/d,

$$\text{Pop} = (9000 \text{ gpm})\left(\frac{1440 \text{ min}}{1 \text{ d}}\right)\left(\frac{1 \text{ person}}{100 \text{ gal/d}}\right)$$
$$\text{Pop} = 13,000 \text{ persons}$$

g. **b.** It needs 4 pumps with 2 spares to meet the daily maximum flows.

h. **a.** The head to pump against will be 40 ft plus the head losses in the system.

$$\text{hp} = \frac{Q\gamma H}{550}$$

$$= \frac{\left(\frac{2500 \text{ gal}}{\text{min}}\right)\left(\frac{1 \text{ ft}^3}{7.48 \text{ gal}}\right)\left(\frac{1 \text{ min}}{60 \text{ s}}\right)\left(\frac{62.4 \text{ lb}}{\text{ft}^3}\right)(40 \text{ ft})}{550 \text{ ft-lb/s-hp}}$$

$$= 25 \text{ hp}$$

i. **d.**

j. **c.** Minimum pressure will occur under the design flow.

7.2 a. **d.** Both bacteria and viruses would contaminate a well. Suspended solids are also of concern to chlorinate the water successfully.

b. **d.** Because of the chance of significant organic pollution in the well, the chlorine dosage has to be large enough to overcome chlorine demand. A dosage of 50 mg/L should be used to provide a safety factor to eliminate the chance of pathogenic contamination.

c. **b.**

$$V = (3.14)\left(\frac{1}{12}\text{ft}\right)^2 (200 \text{ ft})$$
$$V = 4.4 \text{ ft}^3$$

At 50 mg/L of 5% sodium hypochlorite, dosage volume is given in Table 7.11 in *Civil Engineering PE License Review*.

$$\text{SHC} = (4.4 \text{ ft}^3)\left(\frac{28.3 \text{ L}}{\text{ft}^3}\right)\left(\frac{97 \text{ gal}}{100,000 \text{ gal}}\right)$$
$$\text{SHC} = 0.12 \text{ gal}$$

You should always overdose.

d. **d.**

$$\frac{N_t}{N_0} = \frac{1}{1 \times 10^6} = e^{\pm 0.1 t}$$

$$t = 138 \text{ min}$$

e. c.

f. d. The chlorination coefficient, *n*, is unknown.

g. d. Carbonate will affect pH, and iron and manganese can exert a chlorine demand.

h. c. (a) and (b) are notoriously unstable; (d) would be hard to apply.

i. d.

j. a. Sodium hypochlorite is a salt from a weak acid, so the pH would be raised.

7.3 a. b. Using an overflow rate of 50 m/d for alum, or 1230 gal/ft^2, the required area is

$$A = \left(\frac{22.5 \times 10^6 \text{ gal}}{d}\right)\left(\frac{ft^2}{1230 \text{ gal}}\right)$$

$$A = 18,342 \text{ ft}^2$$

b. a.

c. b.

d. c.

e. b.

$$A = \left(\frac{22.5 \times 10^6 \text{ gal}}{d}\right)\left(\frac{1 \text{ ft}^2}{4 \text{ gpm}}\right)\left(\frac{1 \text{ d}}{1440 \text{ min}}\right)$$

$$A = 533 \text{ ft}^2$$

f. d.

g. d.

h. e. The design of unit processes is based upon average daily flow.

i. b.

j. d.

REFERENCES

1. American Water Works Association. *Introduction to Water Distribution: Principles and Practices of Water Supply Operations.* AWWA, Denver, CO, 1986.
2. American Water Works Association. *Water Quality and Treatment.* McGraw-Hill, New York, 1990.
3. American Society of Civil Engineers and the American Water Works Association. *Water Treatment Plant Design.* McGraw-Hill, New York, 1991.
4. Hammer, M. *Water and Wastewater Technology.* John Wiley & Sons, New York, 1975.
5. James M. Montgomery Consulting Engineers. *Water Treatment Principles and Design.* John Wiley & Sons, New York, 1985.
6. McGhee, T. J. *Water Supply and Sewerage.* McGraw-Hill, New York, 1991.
7. Sawyer, C. N., and McCarty, P. L. *Chemistry for Environmental Engineers.* McGraw-Hill, New York, 1978.
8. Steel, E. W., and McCarty, T. J. *Water Supply and Sewerage.* McGraw-Hill, New York, 1979.
9. Tchobanoglous, G., and Schroeder, E. D. *Water Quality.* Addison-Wesley, Menlo Park, CA, 1985.
10. Viessman, W., Jr., and Hammer, M.J. *Water Supply and Pollution Control.* Harper & Row, New York, 1985.

CHAPTER 8

Wastewater Treatment

PROBLEMS

8.1 A schematic of a wastewater treatment plant is shown in Exhibit 8.1.

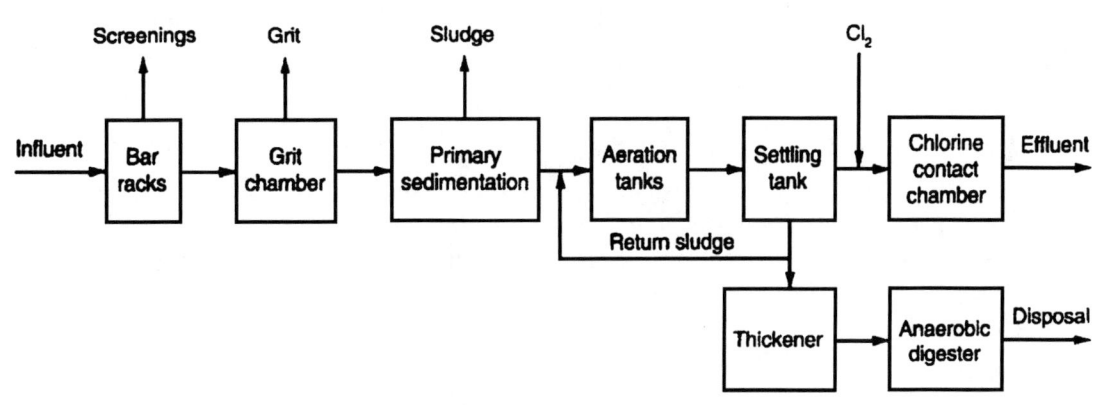

Exhibit 8.1

    a. Treatment to remove grit from the wastewater
        a. decreases BOD
        b. is used to decrease mechanical failure
        c. involves the use of coagulants
        d. can come at any point in the treatment train

    b. The primary settling results in
        a. 50% removal of TSS
        b. sludge that is 1% solids
        c. cost-effective removal of suspended solids by gravity
        d. 25% removal of BOD

    c. A major goal of the biological process is
        a. to remove influent TSS
        b. to convert influent BOD to bacterial cells
        c. to decrease hydraulic detention times
        d. to decrease mechanical failure

d. For sewage treatment plants, the biological treatment unit is often some form of activated sludge process because
  a. activated sludge is easier to operate than other biological treatment units
  b. activated sludge has a greater probability of meeting secondary treatment standards
  c. nitrification can be controlled more easily in activated sludge plants
  d. (b) and (c)

e. A small pinpoint floc is observed in the final clarifier. The best corrective action for this problem would be to
  a. increase the recirculation rate
  b. increase the aeration rate
  c. decrease the food-to-microorganism (F/M) ratio
  d. decrease the solids retention time

f. The sludge is wasted from the biological treatment system to
  a. remove bacteria grown in the biological treatment unit
  b. remove the TSS that are captured in the biological treatment unit
  c. control the solids retention time
  d. All of the above

g. Flotation-thickening processes
  a. are designed around hydraulic detention time
  b. concentrate primary and secondary sludges to about 5% to 10% solids
  c. reduce the size of the sludge processing facilities
  d. result in a wastewater effluent that is less than 30 mg/L of TSS

h. The settling in a secondary clarifier
  a. is designed around hydraulic detention time
  b. involves Type 2 and Type 3 settling
  c. cannot have a recycle flow greater than the plant influent flow
  d. results in a decreased hydraulic detention time in the biological treatment process because of the recycle flow

i. The chlorination of secondary, treated wastewater
  a. eliminates all pathogens in the effluent
  b. reduces effluent BOD
  c. produces only nontoxic products
  d. improves the water quality of the stream

8.2 A two-stage anaerobic digester is used to treat a combined primary and secondary sludge.

a. The second stage of the system is designed to
  a. complete the digestion process
  b. reduce the COD
  c. thicken the sludge
  d. remove BOD

b. The solids retention time in the first state is about
   a. 12 hours
   b. 5 days
   c. 20 days
   d. 35 days

c. The digester volume required per 1 MG of wastewater treated is approximately
   a. 1 ft^3
   b. 3 ft^3
   c. 10 ft^3
   d. 20 ft^3

d. The hydraulic detention time is
   a. equal to the solids retention time
   b. less than the solids retention time
   c. greater than the solids retention time
   d. not an important design variable

e. The pH of the digester
   a. should be about 7.0
   b. can be controlled by the addition of bicarbonate
   c. is affected by carbon dioxide in off-gas
   d. All of the above

f. The temperature in the digester should be controlled to near
   a. 15°C
   b. 25°C
   c. 35°C
   d. 45°C

g. The digestion of primary sludge, as compared to waste-activated sludge,
   a. produces more methane per pound of dry solids
   b. has higher removal of volatile solids
   c. results in higher solids content after treatment
   d. All of the above

h. The heating value of the methane is approximately
   a. 900 BTU/ft^3
   b. 1500 BTU/ft^3
   c. 2000 BTU/ft^3
   d. 2500 BTU/ft^3

i. Compared to aerobic digestion, anaerobic digestion
   a. produces more bacterial cells
   b. uses more energy
   c. is more expensive
   d. is more complex to operate

8.3 A state park is being designed to serve 100 cabins with an average of four persons per cabin. Determine the size of the aerobic stabilization pond needed to treat the expected wastewater flow rate.

8.4 A trickling filter is to be used for treating a domestic sewage flow of 1 MGD, with a BOD_5 concentration after primary sedimentation of 150 mg/L. The recirculation ratio is 2.0. Determine the size of two filters in series to reduce the waste to 20 mg/L.

8.5 A sedimentation basin has an overflow rate of 3 ft/hr. Particle distribution of the influent wastewater is

Percent of Particles	Settling Velocity, ft/hr
20	1 to 2
30	2 to 3
50	3 to 4

Determine the percentage of removal in the sedimentation basin.

8.6 An oxygen transfer test was conducted in pure water with the following results:

Time, min	Dissolved Oxygen, mg/L
0	1.00
0.25	2.00
0.50	2.95
0.75	3.88
1.00	4.65
1.50	5.75
2.00	6.65
2.50	7.30
3.00	7.82
3.50	8.20
4.00	8.55

The test is repeated under steady-state conditions, using an equivalent volume of the wastewater to be tested. Other test conditions remain the same. Test results are

Oxygen saturation concentration for pure water = 9.51 mg/L

Oxygen saturation concentration for wastewater = 9.15 mg/L

Oxygen equilibrium concentration = 5.85 mg/L

Oxygen uptake rate = 51.2 mg/L/hr

a. Calculate the oxygen transfer coefficient K_{La} for the tap water and the wastewater.

b. Calculate the value of α and β.

8.7 A waste has a total dissolved solids concentration of 350 mg/L and a total volatile solids (TVS) concentration of 250 mg/L. The COD is measured as 380 mg/L, and a chemical analysis of the organic compounds gives an empirical formula of CH_2O. The BOD_5 is measured as 220. Assume the BOD decay coefficient is 0.10/d.

a. Determine the theoretical oxygen demand.

b. Determine the ultimate BOD.

c. Determine the degradability of the waste.

8.8 An aeration equipment manufacturer has recommended installation of a mechanical surface aerator with a clean water transfer rate of 3.0 lb O_2/hp-hr. The following additional information is also available:

Dissolved oxygen required in the effluent = 2.0 mg/L

Minimum wastewater temperature = 24°C

Maximum wastewater temperature = 35°C

Plant elevation = 3000 feet above sea level

$\alpha = 0.9$

$\beta = 0.95$

The oxygen uptake rate has been determined as 28 mg/L-hr. The aeration basin volume is 4.1 million gallons. Calculate the annual cost of operation if electricity is 5¢ per kWh.

8.9 An activated sludge process with a flow of 3 MGD has an influent phosphorus concentration of 10 mg/L. The effluent concentration cannot exceed 1 mg/L. Determine the dosage of alum required to achieve the phosphorus removal.

SOLUTIONS

8.1 a. **b.** Grit removal is primarily practiced to keep grit from getting into the bearings in pumps and other mechanical equipment. Another important reason is that the grit will tend to settle into various reactors in the plant.

  b. **c.** Primary sedimentation is the most cost-effective treatment method based on dollars per lb of BOD, or dollars per TSS removed.

  c. **b.** Whereas the biological treatment unit does remove some influent TSS, its main objective is to convert degradable organic compounds into bacterial cells that can be subsequently removed in the secondary clarifier.

  d. **d.** Activated sludge treatment, in general, is more expensive, more complicated, and uses more oxygen than other biological treatment systems. However, it does produce a better settling sludge than most other systems and, as a result, gives a greater probability of meeting a 30 mg/L effluent concentration for both BOD and TSS. It also allows greater control to achieve nitrification.

136 Chapter 8 Wastewater Treatment

e. **d.** Pinpoint floc is typically caused by excess retention of solids or its equivalent, a food/microorganism ratio that is too low.

f. **d.** The waste sludge is comprised of both influent TSS and bacterial growth on the removed BOD. The waste-rate determines the solids retention time.

g. **b.** The objective of sludge-thickening unit operations is to concentrate the waste sludge to reduce the volume of sludge (like anaerobic digestion) in treatment facilities.

h. **b.**

i. **d.** Chlorination is used to reduce the concentration of the bacteria discharged to streams.

8.2 a. **c.**

b. **c.**

c. **b.**

$$V = \left(\frac{300\,\text{mg}}{\text{L}}\right)\left(\frac{150\,\text{gal}}{\text{d}}\right)\left(\frac{3.78\,\text{L}}{\text{gal}}\right)\left(\frac{1\,\text{lb}}{454\times10^3\,\text{mg}}\right)\left(\frac{1000\,\text{ft}^3\cdot\text{d}}{150\,\text{lb}}\right) = 2.5\,\text{ft}^3$$

d. **a.**

e. **d.**

f. **c.**

g. **d.**

h. **a.**

i. **d.**

8.3 The population to be served, P, is

$$P = (100\,\text{cabins})\left(\frac{4\,\text{persons}}{\text{cabin}}\right) = 400\,\text{persons}$$

Assuming a 50 gal/d-capita usage, the daily discharge, Q, is

$$Q = \left(\frac{50\,\text{gal}}{\text{d-capita}}\right)(400\,\text{persons}) = 2.0\times10\,\text{gpd}$$

Using a factor of 0.17 lb BOD_5/d-capita, the BOD_5 loading is

$$BOD_5 = \left(\frac{0.17\,\text{lb BOD}_5}{\text{d}}\right)(400\,\text{persons}) = 68\,\text{lb BOD}_5/\text{d}$$

Base the design on 100 lb BOD_5/acre-d and a 40-d detention time:

$$A = \left(\frac{68 \text{ lb BOD}_5}{d}\right)\left(\frac{1 \text{ acre-d}}{100 \text{ lb}}\right) = 0.68 \text{ acre; use } 0.8 \text{ acre}$$

$$V = \left(\frac{2.0 \times 10^4 \text{ gal}}{d}\right)(40d)\left(\frac{1 \text{ ft}^3}{7.48 \text{ gal}}\right) = 1.1 \times 10^5 \text{ ft}^3$$

$$D = \left(\frac{1.1 \times 10^5 \text{ ft}^3}{0.8 \text{ acres}}\right)\left(\frac{1 \text{ acre}}{43{,}560 \text{ ft}^2}\right) = 3.1 \text{ ft; use 3 feet}$$

8.4

$$E_{design} = (130/150)(100) = 87\%$$

$$F = \frac{1+R}{(1+R_{10})^2} = \frac{1+2}{\left(1+\frac{2}{10}\right)^2} = 2.08$$

Assume the same volume for each filter:

$$BOD = \left(\frac{150-20 \text{ mg}}{L}\right)\left(\frac{1 \times 10^6 \text{ gal}}{d}\right)\left(\frac{1 \text{ lb}}{454 \times 10^3 \text{ mg}}\right)\left(\frac{3.78 \text{ L}}{\text{gal}}\right) = 1.1 \times 10^3 \text{ lb/d}$$

Now apply Eq. (8.36) of *Civil Engineering PE License Review* to each filter:

$$E_1 = \frac{100}{1+0.0561\sqrt{\frac{1100/V}{2.08}}} \quad E_2 = \frac{100}{1+\frac{0.0561}{1-\frac{E_1}{100}}\sqrt{\frac{1100/V}{2.08}}} \quad E = 87 = E_1 + (100-E_1)(E_2)$$

Trial and error

V	E_1	E_2	$E_{overall}$
$18.5 \times 10^3 \text{ ft}^3$	70	56	$70 + (1-0.70)(56) = 87\%$

8.5

Avg. Settling Velocity, ft/hr	% Particles	% Removed	% Remaining
1.5	20	$\frac{1.5}{3.0} \times 100 = 50$	10
2.5	30	$\frac{2.5}{3.0} \times 100 = 83$	5
3.5	50	100	0
			Total = 15%

The percentage removal is 85 percent.

8.6 a. For non-steady-state conditions, the rate of oxygen transfer is given by the equation

$$\frac{dC}{dt} = K_{La}(C_s - C)$$

The equation may be integrated and rewritten as

$$K_{La} = \frac{2.303 \log_{10}[(C_s - C_1)/(C_s - C_2)]}{t_2 - t_1}$$

where K_{La} = transfer coefficient, C_s = saturation concentration in mg/L, and C_1, C_2 = dissolved oxygen concentration at times t_1 and t_2.

K_{La} is calculated as

$$K_{La} = \frac{2.303 \log_{10}[(9.51 - 1.0)/(9.51 - 8.20)]}{3.5 \text{ min}}$$

$$K_{La} = 0.53 \text{ min}^{-1}$$

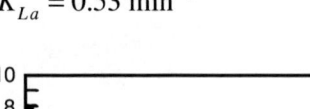

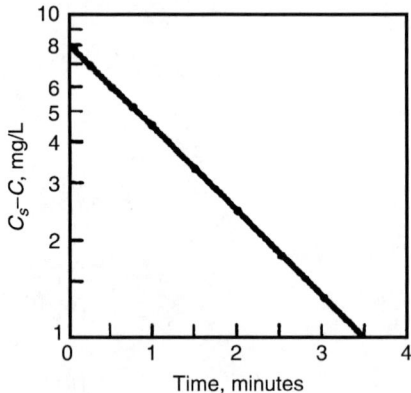

Exhibit 8.6

For steady-state conditions, K_{La} may be calculated from the equation

$$\frac{dC}{dt} = r - K_{La}(C_s - C)$$

At steady state $dC/dt = 0$, the equation may be written as

$$K_{La} = \frac{r}{C_s - C_L}$$

where r = oxygen uptake rate in mg/L/hr, C_s = oxygen saturation concentration in mg/L, and C_L = oxygen equilibrium concentration in mg/L.

$$K_{La} = \frac{51.2}{9.15 - 5.85} = 15.5 \text{ h}^{-1} = 0.258 \text{ min}^{-1}$$

b. $$\alpha = \frac{K_{La} \text{ waste water}}{K_{La} \text{ tap water}}$$

$$\alpha = 0.258/0.530 = 0.487$$

$$\beta = \frac{\text{oxygen saturation waste water}}{\text{oxygen saturation tap water}}$$

$$\beta = 9.15/9.51 = 0.962$$

8.7 a. The ThOD equation is

$$CH_2O + \frac{3}{2}O_2 = CO_2 + 2H_2O$$

$$\text{ThOD} = \left(\frac{250 \text{ mg TVS}}{L}\right)\left(\frac{1 \text{ M CH}_2O}{30 \text{ g TVS}}\right)\left(\frac{1.5 \text{ M O}_2}{1 \text{ M CH}_2O}\right)\left(\frac{32 \text{ g O}_2}{1 \text{ M O}_2}\right) = 400 \text{ mg/L}$$

b. $$\text{BOD}_L = \frac{\text{BOD}_5}{(1-10^{-kt})} = \frac{220}{[1-10^{-0.1(5)}]} = 322 \text{ mg/L}$$

c. $$\frac{\text{BOD}_L}{\text{COD}} = \frac{322 \text{ mg/L}}{380 \text{ mg/L}} = 0.85$$

A waste with $\frac{\text{BOD}_L}{\text{COD}} = 0.8$ is considered to be highly degradable.

8.8 Calculate field transfer rate, N, using the equation

$$N = N_0 \frac{\beta \times C_{sat} - C_l}{C_{sc}} 1.024^{T-20} \alpha$$

where N_0 = clean water transfer rate, C_{sat} = saturation concentration at design temperature and altitude = 6.4 mg/L, C_l = dissolved oxygen level of aeration basin, C_{sc} = saturation concentration at sea level and 20 °C = 9.2 mg/L, and T = wastewater temperature in °C.

Assume the dissolved oxygen of the aeration tank equals the dissolved oxygen required in the effluent. The maximum wastewater temperature is used for design.

$$N = 3.0 \frac{0.95 \times 6.4 - 2.0}{9.2} \times 1.024^{35-20} \times 0.9 = 1.71 \text{ lb O}_2/\text{hp-hr}$$

Calculate the pounds of oxygen required:

4.1 million gallons × 28 mg/L-hr × 24 hr/day × 8.34 = 23,000 lb/day

The horsepower required is

$$\frac{23,000 \text{ lb/day}}{24 \times 1.75 \text{ lb O}_2/\text{hp-hr}} = 548 \text{ hp}$$

Hence, the cost of 365 days/year operation and 94% motor efficiency (assumed) is

$$\frac{548 \text{ hp}}{0.94} \times \frac{0.746 \text{ kW}}{1 \text{ hp}} \times \frac{\$0.05}{\text{kWh}} \times \frac{24 \text{ hr}}{1 \text{ d}} \times \frac{365 \text{ d}}{1 \text{ yr}} = \$190,500/\text{year}$$

8.9 The overall equation is

$$Al^{3+} + PO_4^{3-} \rightarrow AlPO_4$$

$$\text{Theoretical Al} = (10 \text{ mg P/L} \pm 1 \text{ mg P/L}) \cdot \frac{1 \text{ M } PO_4^{3-}}{31 \text{ g-P}} \cdot \frac{1 \text{ M Al}}{1 \text{ M } PO_4^{3-}} \cdot \frac{342 \text{ g Al}}{1 \text{ M Al}}$$

$$= 99 \text{ mg Al/L}$$

$$\text{Actual Al} = 2.3 \times (\text{Theoretical Al}) = 230 \text{ mg Al/L}$$

REFERENCES

1. American Society of Civil Engineering. *Design and Construction of Sanitary and Storm Sewers*. ASCE, New York, 1969.
2. Hammer, M. *Water and Wastewater Technology*. John Wiley & Sons, New York, 1975.
3. McGhee, T. J. *Water Supply and Sewerage*. McGraw-Hill, New York, 1991.
4. Metcalf and Eddy, Inc. *Wastewater Engineering: Treatment, Disposal, and Reuse*. McGraw-Hill, New York, 1991.
5. Sawyer, C. N., and McCarty, P. L. *Chemistry for Environmental Engineers*. McGraw-Hill, New York, 1978.
6. Steel, E. W., and McGhee, T. J. *Water Supply and Sewerage*. McGraw-Hill, New York, 1979.
7. Tchobanoglous, G., and Schroeder, E. D. *Water Quality*. Addison-Wesley, Menlo Park, CA, 1985.
8. Viessman, W., Jr., and Hammer, M. J. *Water Supply and Pollution Control*. Harper & Row, New York, 1985.
9. Water Environment Federation and American Society of Civil Engineers. *Design of Municipal Wastewater Treatment Plants: Vol 1*. Book Press, Brattleboro, VT, 1991.
10. Water Environment Federation and American Society of Civil Engineers. *Design of Municipal Wastewater Treatment Plants: Vol 2*. Book Press, Brattleboro, VT, 1991.

CHAPTER 9

Geotechnical Engineering

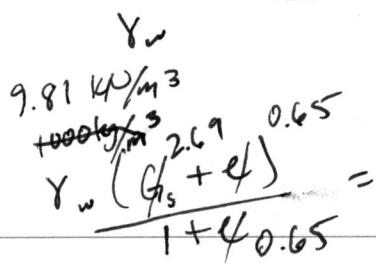

PROBLEMS

9.1 A saturated soil has a void ratio $e = 0.65$ and a specific gravity of soil solids $G_s = 2.69$. The unit weight (saturated) of this soil is most nearly:
a. 16.5 kN/m³
b. 17.3 kN/m³
c. 18.1 kN/m³
d. 19.7 kN/m³

9.2 Following are the results of a standard Proctor compaction test.

Moisture Content, w (%)	Weight of Moist Soil in Proctor Mold, W (lb)
10	3.30
12	3.57
14	3.77
16	3.90
18	3.87
20	3.67

The maximum dry unit weight of the soil is most nearly:
a. 100 lb/ft³
b. 106 lb/ft³
c. 112 lb/ft³
d. 115 lb/ft³

9.3 The specific gravity of solids, G_s, for a soil is 2.75. The zero-air-void dry unit weight of the soil at a moisture content of 20 percent is about
a. 17.5 kN/m³
b. 18.1 kN/m³
c. 18.9 kN/m³
d. 19.7 kN/m³

141

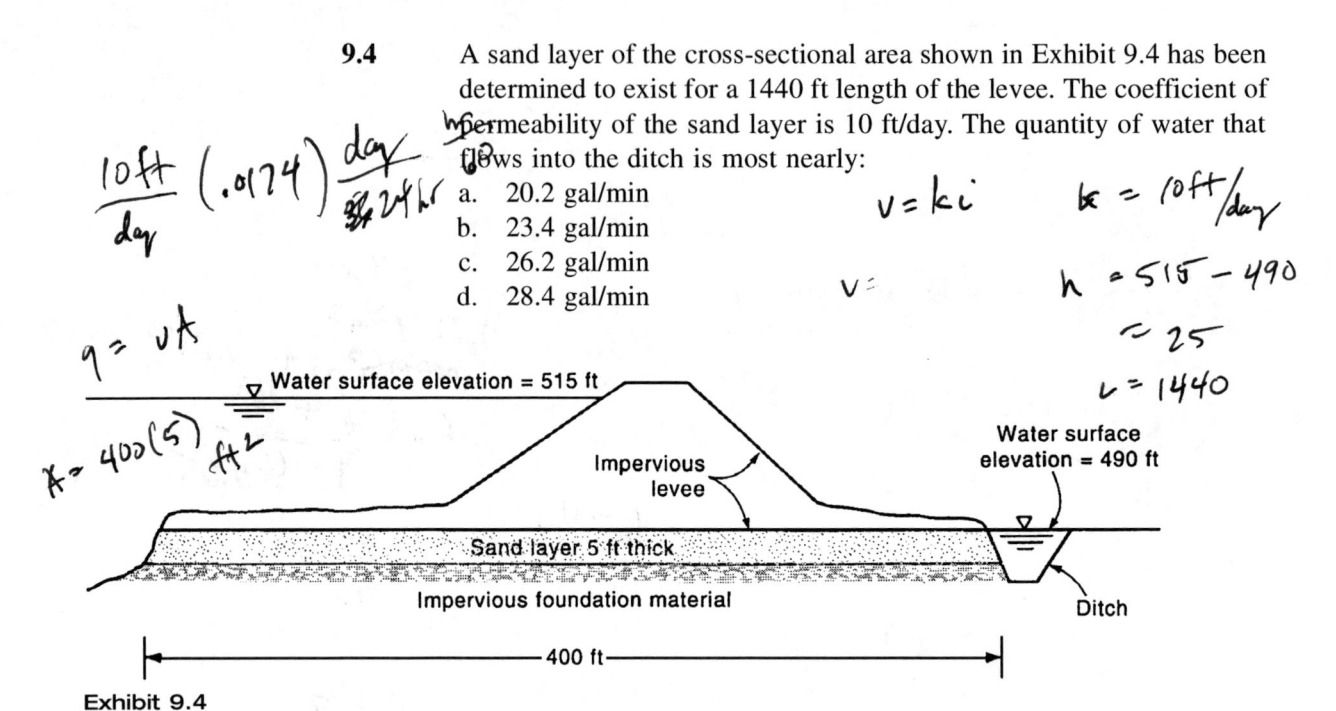

9.4 A sand layer of the cross-sectional area shown in Exhibit 9.4 has been determined to exist for a 1440 ft length of the levee. The coefficient of permeability of the sand layer is 10 ft/day. The quantity of water that flows into the ditch is most nearly:

a. 20.2 gal/min
b. 23.4 gal/min
c. 26.2 gal/min
d. 28.4 gal/min

Exhibit 9.4

9.5 An earth dam on a pervious but strong earth foundation has the cross section shown in Exhibit 9.5. The core of the dam is sealed from the jointed rock foundation with a thin layer of gunite or slush grout. The seepage through the foundation of the dam (in ft^3/day/ft of the dam) is nearly equal to

a. 2
b. 2.5
c. 3
d. 3.75

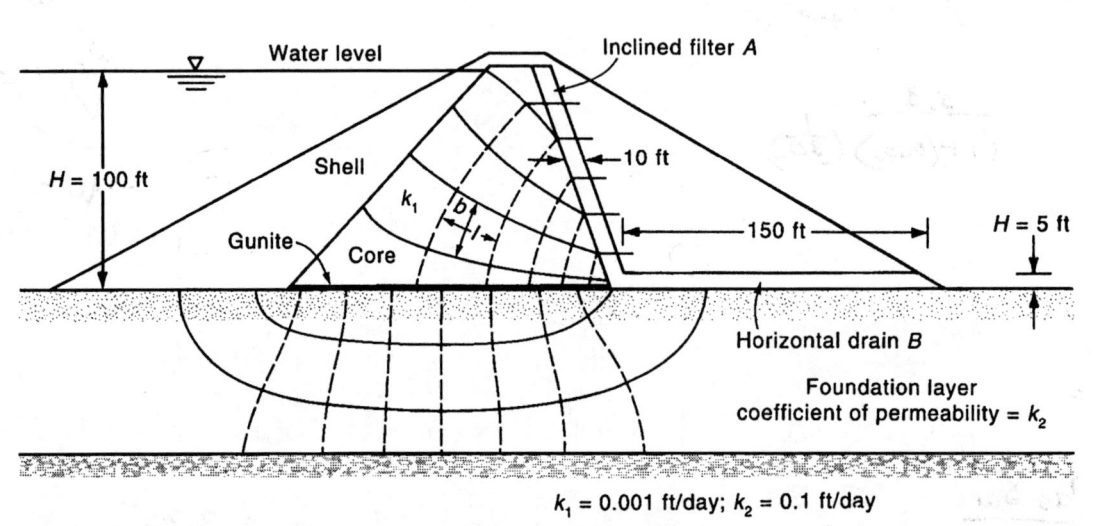

Exhibit 9.5

9.6 Refer to the flow net through the earth dam shown in Exhibit 9.5. The seepage through the dam only (ft³/day/ft of dam) is about
 a. 0.055
 b. 0.15
 c. 0.2
 d. 0.3

9.7 For a large construction project, the general subsoil condition is shown in Exhibit 9.7. If a 10-foot-thick fill is placed over the existing ground surface, the consolidation settlement of the soft clay layer (unit weight of fill = 110 lb/ft³) will be most nearly:
 a. 2 in.
 b. 3 in.
 c. 4 in.
 d. 5 in.

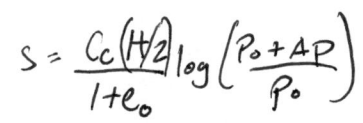

$S = \dfrac{C_c(H/2)}{1+e_o} \log\left(\dfrac{P_o+\Delta P}{P_o}\right)$

$C_c = 0.009(42-10) = .288$

$C_s = \tfrac{1}{5} C_c = .0576$

$P_o = 100(0) + (100 - 0)5$
$+ \cancel{100}(120-62.4)5 +$
$118 - 62.4(5) =$
$2,066 \ \dfrac{lb}{ft^2} \times \dfrac{ft^2}{12^2 \, in^2}$

$\Delta P =$

10' $\gamma = 100 \ \dfrac{lb}{ft^3}$

Original ground surface

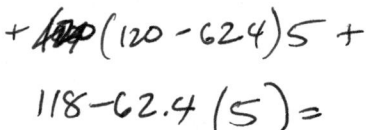

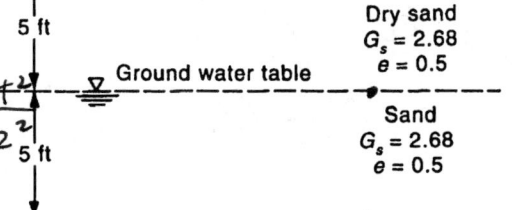

5 ft — Dry sand, $G_s = 2.68$, $e = 0.5$

Ground water table

5 ft — Sand, $G_s = 2.68$, $e = 0.5$

10 ft — Soft, normally consolidated clay, $G_s = 2.72$, $e = 0.92$, Liquid limit = 42%

Sand

Exhibit 9.7

118

$\dfrac{e_o}{G_s} = w$

$\dfrac{.92}{2.72} = .3382$

$\gamma_{sat} = \dfrac{62.4(2.72+.92)}{1+.92}$

9.8 A 25-mm-thick clay specimen was subjected to a conventional consolidation test. When the effective pressure on the soil specimen was increased from 96 kN/m² to 192 kN/m², the void ratio decreased from 0.75 to 0.62. Also, the time for 50 percent consolidation was determined to be 3 min. The coefficient of permeability, k, of the clay for the loading range is nearly equal to
 a. 7×10^{-8} m/min
 b. 8×10^{-8} m/min
 c. 9×10^{-8} m/min
 d. 10×10^{-8} m/min

• 96 kN/m²   192 kN/m²

$k = ?$

$v = ki$

$q = kiA$

STOP

9.9 Laboratory tests on a 25 mm thick clay sample drained at the top and bottom show that 50 percent consolidation takes place in 11 minutes. The time for a similar clay layer in the field, 4 m thick and drained at the top only, to undergo 50 percent consolidation is most nearly:
a. 400 days
b. 500 days
c. 600 days
d. 780 days

9.10 An embankment will be constructed from borrowed soil. The cross section of the embankment is shown in Exhibit 9.10. The soil in the embankment has to be compacted to a dry unit weight of 108 lb/ft^3 at a moisture content of 14.5 percent. The borrowed soil will be brought in 10-cubic-yard truck loads. The borrowed soil has a void ratio of 1.1. Given: specific gravity of soil solids, $G_s = 2.68$. The volume of excavation needed at the borrow site for construction of each 100 ft length of embankment is most nearly:

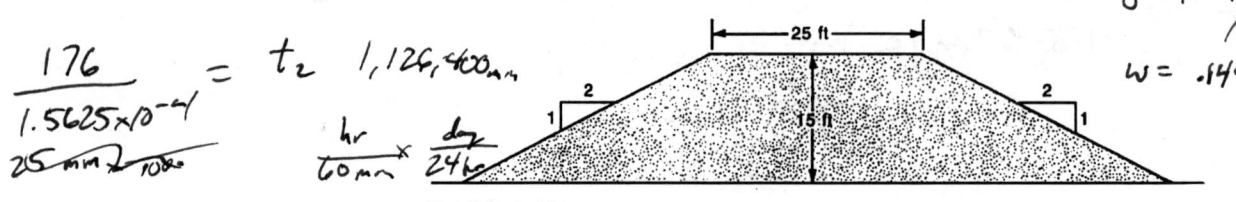

Exhibit 9.10

a. 4144 yd^3
b. 5260 yd^3
c. 3550 yd^3
d. 4890 yd^3

9.11 A retaining wall is shown in Exhibit 9.11. The Rankine active earth pressure, σ_a, at B is most nearly:

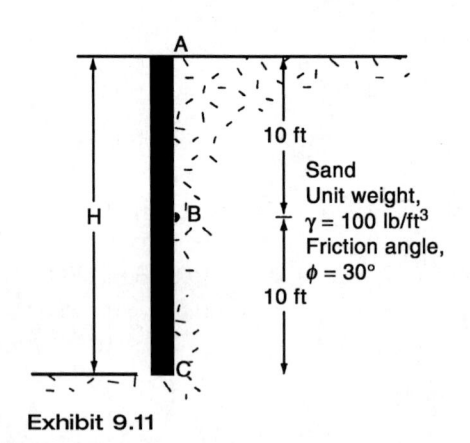

Exhibit 9.11

a. 230 lb/ft^2
b. 330 lb/ft^2
c. 425 lb/ft^2
d. 475 lb/ft^2

9.12 Refer to the retaining wall shown in Exhibit 9.11. The Rankine active force, P_a, per unit length of the wall will be nearly:
a. 3600 lb/ft
b. 5400 lb/ft
c. 6600 lb/ft
d. 9000 lb/ft

9.13 A sheet pile bulkhead is driven along a shoreline as shown in Exhibit 9.13. The bottom outside the piling is dredged, and a sand fill is placed behind the bulkhead as shown. Neglect settlement during fill placement. Use the expression

$$p'_a = p'_v \tan^2\left(45 - \frac{\phi}{2}\right) - 2c \tan\left(45 - \frac{\phi}{2}\right)$$

where

p'_a = active horizontial pressure
p'_v = effective vertical pressure

Assume $G_s = 2.65$. The initial active force acting on the land side of the sheet piling is nearly equal to
a. 6000 lb/ft
b. 8000 lb/ft
c. 11,000 lb/ft
d. 13,000 lb/ft

	Dry unit weight, γ_d, lb/ft³	Water content, w, %	Cohesion c lb/ft²	Soil function angle, ϕ, degrees
Medium sand fill El. 0 ft	100	10.0	0	28
Natural medium-to-coarse sand El. −12 ft	100	24.5	0	28
Sandy clay El. −18 ft	83	37.0	275	16
Coarse sand	105	21.6	0	36

Exhibit 9.13

9.14 A six-story steel frame building is to be underlain by two levels of underground parking. The necessary excavation will extend 30 ft below the adjacent streets and ground level, with the subgrade level beneath the lowest floor slab at El. 70 ft, as shown in Exhibit 9.14. The soil is silty clay and has the following properties:

Cohesion, $c = 540$ lb/ft^2
Angle of internal friction, $\varphi = 10°$
In situ unit weight $= 125$ lb/ft^3
Active soil pressure (assumed for design) $= 20$ lb/ft^3 equivalent
Passive soil pressure (assumed for design) $= 1000$ lb/ft^3 equivalent

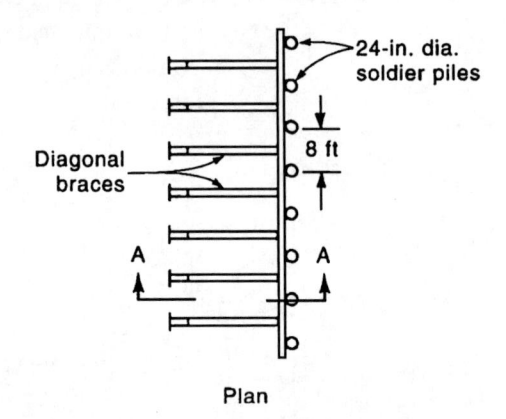

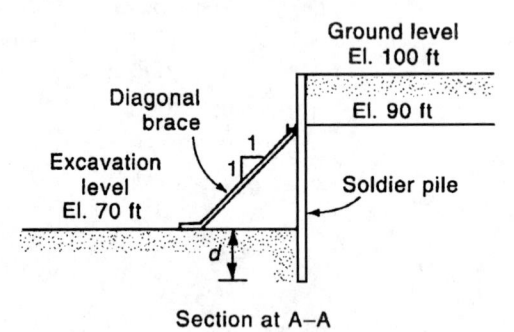

Exhibit 9.14

One wall of the underground parking area will be constructed immediately adjacent (within a few inches) of the property line. Assume the groundwater level is well below the excavated depth. Twenty-four-inch-diameter concrete soldier piles, on 8-foot centers, will be installed to maintain the vertical embankment. For stability, d must be nearly equal to:

a. 2 ft
b. 3 ft
c. 3.5 ft
d. 4 ft

9.15 Following are the results of four drained direct shear tests on a normally consolidated clay. The sample has a diameter of 50 mm and a height of 25 mm.

Test No.	Normal Force, N	Shear Force at Failure, N
1	271	120.6
2	406.25	170.64
3	474	204.1
4	541.65	244.3

The average drained angle of friction is about:
a. 18°
b. 20°
c. 24°
d. 28°

9.16 For a normally consolidated clay, the results of a drained triaxial test are as follows:

chamber confining pressure = 140 kN/m^2
deviator stress at failure = 263.5 kN/m^2

The soil friction angle, ϕ, is nearly equal to
a. 20°
b. 25°
c. 29°
d. 32°

9.17 A 30 ft high slope is shown in Exhibit 9.17. For the slope, the unit weight of soil, γ, is 108 lb/ft^3. Along the soil-rock interface, the friction angle, ϕ, is 20°, and the cohesion, c, is 500 lb/ft^2. For the wedge ABC, the factor of safety against sliding along the rock surface is nearly equal to:

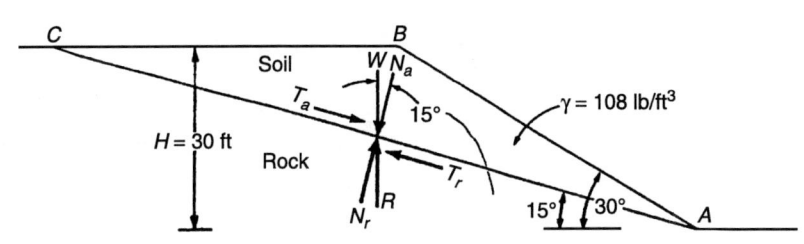

Exhibit 9.17

a. 1.5
b. 2.6
c. 3.7
d. 4.8

9.18 A saturated clay slope is shown in Exhibit 9.18. The factor of safety against sliding along the circular surface $\overline{AC}$ is nearly equal to
a. 1.1
b. 1.8
c. 2.3
d. 2.8

Note: The weight of ABC per unit length of the slope is 32.4 kip/ft.

148 Chapter 9 Geotechnical Engineering

Exhibit 9.18

9.19 The cross section of a cantilever retaining wall is shown in Exhibit 9.19. Assume the following: weight of concrete = 150 lb/ft^3; friction angle between concrete and base soil, $\delta = 15°$; adhesion between concrete and base soil, $c_a = 300$ lb/ft^2. The factor of safety against overturning is nearly equal to:

a. 2.1
b. 3.2
c. 4.1
d. 5.1

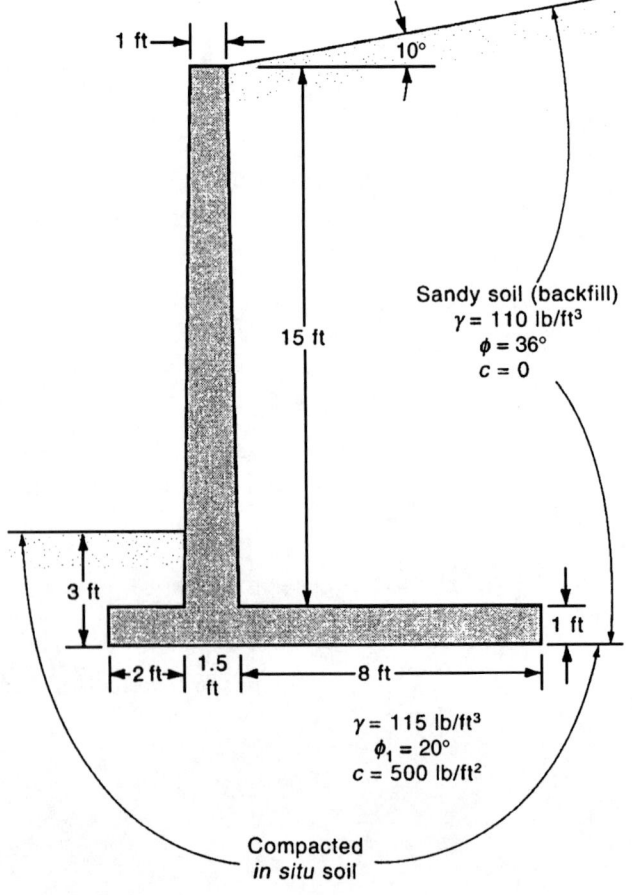

Exhibit 9.19

9.20 Refer to the retaining wall described in Exhibit 9.19. For this wall, the factor of safety with respect to sliding is about
a. 1.5
b. 2.5
c. 3.2
d. 3.9

SOLUTIONS

9.1 **d.** Saturated unit weight is given by

$$\gamma_{sat} = \frac{\gamma_w(G_s + e)}{1+e} = \frac{(9.81)(2.69+0.65)}{1+0.65} = 19.9 \text{ kN/m}^3$$

9.2 **a.** The volume of a standard Proctor mold is $\frac{1}{30}$ ft^3. Also

$$\gamma_d = \frac{\gamma}{1+w}$$

Now the following table can be prepared.

w(%)	W(lb)	γ $W \times 30$ (lb/ft^3)	γ_d (lb/ft^3)
10	3.30	99	90
12	3.57	107.1	95.63
14	3.77	113.1	99.21
16	3.90	117.0	100.86
18	3.87	116.1	98.39
20	3.67	110.1	91.75

A plot of γ_d versus w is shown in Exhibit 9.2. From this, the maximum dry unit weight = 101 lb/ft^3.

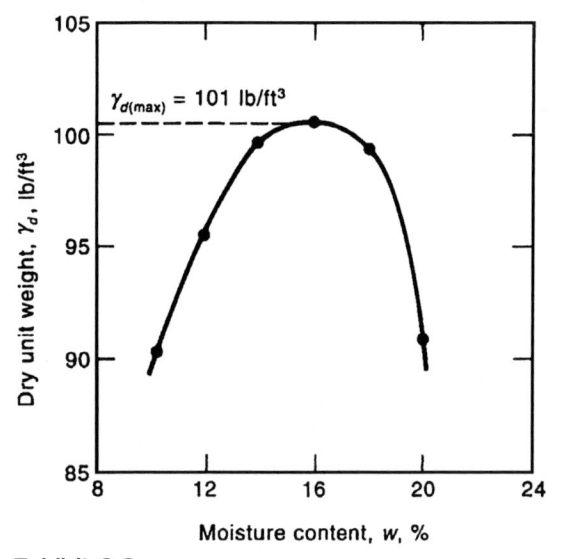

Exhibit 9.2

9.3 a.

$$\gamma_{zav} = \frac{\gamma_w}{w + \frac{1}{G_s}} = \frac{9.81}{0.2 + \frac{1}{2.75}} = 17.4 \text{ kN/m}^3$$

9.4 b. According to Darcy's law,

$$q = kiA$$
$$i = \frac{h}{l} = \frac{515 - 490}{400} = 0.0625$$
$$A = (5)(1440) = 7200 \text{ ft}^2$$

So

$$q = kiA = (10)(0.0625)(7200) = 4500 \text{ ft}^3/\text{day}$$
$$= (4500 \text{ ft}^3/\text{day})(7.48 \text{ gal/ft}^3)\left(\frac{1}{1440} \text{ day/min}\right)$$
$$= 23.4 \text{ gal/min}$$

9.5 d.

$$Q = \frac{N_f}{N_d} k_2 h$$

where

N_f = number of flow channels

N_d = number of equipotential drops

h = total head dissipated

$$Q = \frac{3}{8}(0.1)(100) = 3.75 \text{ ft}^3/\text{day/ft of dam}$$

9.6 a. Flow through each channel

$$\Delta Q = k_1 i b \text{ ft}^3/\text{day/ft}$$

where
$i = \frac{\Delta h}{l}$ = hydraulic gradient
b = normal distance between flow lines

The flow net in the core has four flow channels: No. 1 at the bottom; No. 4 at the top. Thus,

$$Q = k_1 \sum_{j=1}^{4} i_j b_j = k_1 \sum_{j=1}^{4} (\Delta h_j)\left(\frac{b}{l}\right)_j$$

In channel No. 1, the flow net is nearly rectangular, so say $\left(\frac{b}{l}\right)_1 = \frac{1}{2}$; for channels 2–4 say $\left(\frac{b}{l}\right)_j = 1$. Next we want the average head loss h_L across the core in each zone: $\frac{P}{\gamma_w} + Z_l$ = constant = 100 ft on upstream face.

On the downstream face of the core, $\frac{P}{\gamma_w} = 0$ is assumed in the drain. So $Z_l + h_L$ = constant = 100 ft on downstream face. Using the average Z_l for each zone, by scaling we have

Zone 1: $Z_l = 2$ ft; $h_L = 98$ ft; $\Delta h = \frac{h_L}{6} = \frac{98}{6}$

Zone 2: $Z_l = 10$ ft; $h_L = 90$ ft; $\Delta h = \frac{h_L}{5.5} = \frac{90}{5.5}$

Zone 3: $Z_l = 25$ ft; $h_L = 75$ ft; $\Delta h = \frac{h_L}{4.5} = \frac{75}{4.5}$

Zone 4: $Z_l = 55$ ft; $h_L = 45$ ft; $\Delta h = \frac{h_L}{3} = \frac{45}{3}$

$$Q = k_1 \sum_{j=1}^{4} (\Delta h_j)\left(\frac{b}{l}\right) = (0.001)\left[\left(\frac{1}{2}\right)\left(\frac{98}{6}\right) + \frac{90}{5.5} + \frac{75}{4.5} + \frac{45}{3}\right]$$

$Q = 0.056$ ft^3/day/ft of dam

Note: For channel No. 1, $b/l = 1/2$. For channel Nos. 2–4, $b/l = 1$.

9.7 d.

$$S = \frac{C_c H}{1+e_0} \log \frac{p_0 + \Delta p}{p_0}$$

$C_c \approx 0.009(LL - 10) = (0.009)(42 - 10) = 0.288$

$e_0 = 0.92$

$H = 10$ ft $= 120$ in.

$\Delta p =$ (height of fill)(110) = (10)(110) = 1100 lb/ft^2

Calculation of p_0 at the middle of the clay layer:

Dry unit weight of sand:

$$\frac{G_s \gamma_w}{1+e} = \frac{(2.68)(62.4)}{1+0.5} = 111.5 \text{ lb/ft}^3$$

Effective unit weight of sand:

$$\frac{(G_s - 1)\gamma_w}{1+e} = \frac{(2.68-1)(62.4)}{1+0.5} = 69.9 \text{ lb/ft}^3$$

Effective unit weight of clay:

$$\frac{(G_s - 1)\gamma_w}{1+e} = \frac{(2.72-1)(62.4)}{1+0.92} = 55.9 \text{ lb/ft}^3$$

$$p_0 = (5)(111.5) + (5)(69.9) + \left(\frac{10}{2}\right)(55.9) = 557.5 + 349.5 + 279.5 = 1186.5 \text{ lb/ft}^2$$

$$S = \frac{(0.288)(120)}{1+0.92} \log \frac{1186.5 + 1100}{1186.5} = 5.13 \text{ in.}$$

9.8 b. Coefficient of consolidation:

$$c_v = \frac{T_v H_d^2}{t}$$

$$H_d = \text{length of drainage path} = \frac{25 \text{ mm}}{2} = 12.5 \text{ mm}$$

$$t = 3 \text{ min}$$

For 50 percent consolidation, $T_v = 0.197$,

$$c_v = \frac{(0.197)(12.5)^2}{3} = 10.26 \text{ mm}^2/\text{min} = 10.26 \times 10^{-6} \text{ m}^2/\text{min}$$

Volume coefficient of compressibility:

$$m_v = \frac{\frac{\Delta e}{\Delta p}}{1 + e_{av}}$$

$$\Delta e = 0.75 - 0.62 = 0.13$$

$$\Delta p = 192 - 96 = 96 \text{ kN/m}^2$$

$$e_{av} = \frac{0.75 + 0.62}{2} = 0.685$$

$$m_v = \frac{\frac{0.13}{96}}{1 + 0.685} = 80.36 \times 10^{-5} \text{ m}^2/\text{kN}$$

$$k = c_v m_v \gamma_w = (10.26 \times 10^{-6})(80.36 \times 10^{-5})(9.81) = 8.09 \times 10^{-8} \text{ m/min}$$

9.9 d.

$$T_v = \frac{c_v t}{H_{dr}^2}$$

So

$$\frac{t_{lab}}{H_{dr(L)}^2} = \frac{t_F}{H_{dr(F)}^2}$$

Or

$$\frac{11 \text{ min}}{\left(\frac{0.025 \text{ m}}{2}\right)^2} = \frac{t_F}{(4 \text{ m})^2}$$

$$T_F = 1{,}126{,}400 \text{ min} = 782 \text{ days}$$

9.10 a. The volume of the embankment per 100-foot length is

$$V = [\tfrac{1}{2}(15 \times 30) + (25 \times 15) + \tfrac{1}{2}(15 \times 30)](100) = 82{,}500 \text{ ft}^3$$

The dry weight of soil per 100-foot length of embankment is

$$(82{,}500)(108) = 8{,}910{,}000 \text{ lb}$$

The dry unit weight of soil at the borrow site is

$$\gamma_d = \frac{G_s \gamma_w}{1+e} = \frac{(2.68)(62.4)}{1+1.1} = 79.63 \text{ lb/ft}^3$$

The volume of excavation needed is

$$\frac{8,910,000}{79.63} = 111,892.5 \text{ ft}^3 = 4144.2 \text{ yd}^3$$

9.11 b.

$$K_a = \tan^2\left(45 - \frac{\phi}{2}\right) = \tan^2\left(45 - \frac{30}{2}\right) = 0.33$$

At B, $\sigma_a = \gamma z K_a = (100)(10)(0.33) = 330 \text{ lb/ft}^2$

9.12 c.

$$P_a = \tfrac{1}{2}\gamma H^2 K_a = \tfrac{1}{2}(100)(20)^2\left[\tan^2\left(45 - \frac{30}{2}\right)\right] = 6600 \text{ lb/ft}$$

9.13 c. Following are calculations of moist (γ) and effective (γ') unit weights for each layer. Note

$$\gamma = \gamma_d (1 + w)$$

where

γ_d = dry unit weight

w = water content

$\gamma' = \gamma_{sat} - \gamma_w$

$\gamma_w = 62.4 \text{ lb/ft}^3$

Soil	γ_d, lb/ft³	γ, lb/ft³	γ', lb/ft³
Medium sand	100	110	—
Medium-to-coarse sand	100	124.5	62.1
Sandy clay	83	113.7	51.3
Coarse sand	105	127.7	65.3

This problem is subject to various interpretations. Below is one approach.

$$p'_a = p'_v \tan^2\left(45 - \frac{\phi}{2}\right) - 2c\tan\left(45 - \frac{\phi}{2}\right)$$

At El. + 6 ft: $p'_v = 0$, $c = 0$, $p'_a = 0$

At El. 0 ft:

In top layer: $p'_v = (110)(6) = 660 \text{ lb/ft}^2; c = 0$

$$p'_a = 660 \tan^2\left(45 - \frac{28}{2}\right) = 238 \text{ lb/ft}^2$$

In bottom layer: $p'_a = 660 \tan^2\left(45 - \frac{28}{2}\right) = 238 \text{ lb/ft}^2$

At El. − 12 ft:

$$p'_v = (110)(6) + (62.1)(12) = 1405.2 \text{ lb/ft}^2$$

In top layer: $p'_a = 1405.2 \tan^2\left(45 - \frac{28}{2}\right) = 507 \text{ lb/ft}^2$

In bottom layer: $p'_a = 1405.2 \tan^2\left(45 - \frac{16}{2}\right) - (2)(275)\tan\left(45 - \frac{16}{2}\right) = 383 \text{ lb/ft}^2$

At El. − 18 ft:

$$p'_v = 1405.2 + (51.3)(6) = 1713 \text{ lb/ft}^2$$

In top layer: $p'_a = 1713 \tan^2\left(45 - \frac{16}{2}\right) - (2)(275)\tan\left(45 - \frac{16}{2}\right) = 558 \text{ lb/ft}^2$

In bottom layer: $p'_a = 1713 \tan^2\left(45 - \frac{36}{2}\right) = 445 \text{ lb/ft}^2$

At El. − 24 ft:

$$p'_v = 1713 + (65.3)(6) = 2104.8 \text{ lb/ft}^2$$

$$p'_a = 2104.8 \tan^2\left(45 - \frac{36}{2}\right) = 546 \text{ lb/ft}^2$$

The plot of active pressure with depth is shown in Exhibit 9.13a.

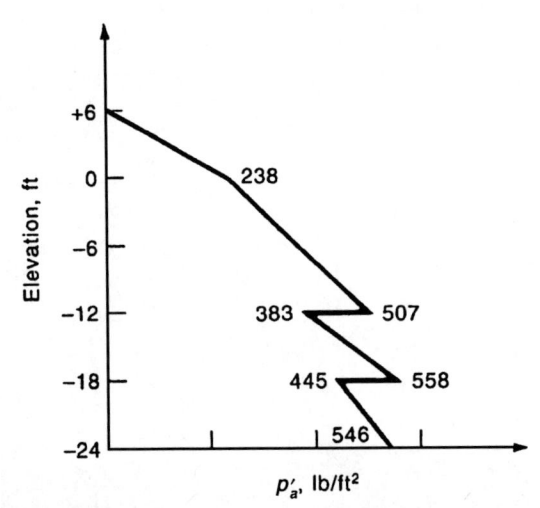

Exhibit 9.13a

The area of the pressure diagram is calculated as follows:

$$P_a = \tfrac{1}{2}(6)(238) + (12)\left(\frac{238+507}{2}\right) + (6)\left(\frac{383+558}{2}\right) + (6)\left(\frac{445+546}{2}\right)$$
$$= 10,980 \text{ lb/ft}$$

9.14 c. Assumptions:

- Since design soil pressures are given in terms of equivalent fluid pressures, use triangular pressure distributions for active and passive pressures.
- The 24 in. diameter soldier piles spaced 8 ft center-to-center are satisfactory to support the silty clay.
- Pressures and forces shown in Exhibit 9.14a are for a 1 ft section normal to the cross section.

For stability, there must be moment equilibrium about point O.

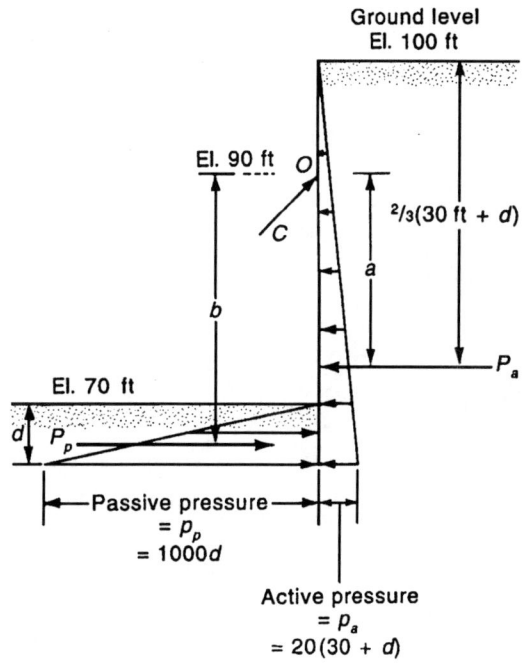

Exhibit 9.14a

$$P_a = \frac{(20)(30+d)(30+d)}{2} = 9000 + 600d + 10d^2 \text{ (lb/ft)}$$

$$P_p = \frac{1000d^2}{2} = 500d^2$$

$$a = \tfrac{2}{3}(30+d) - 10 = 10 + \tfrac{2}{3}d$$

$$b = 20 + \tfrac{2}{3}d$$

$$\Sigma M_O = 0$$

$$P_a \times a - P_p \times b = 0$$

$$(9000 + 600d + 10d^2)(10 + \tfrac{2}{3}d) = (500d^2)(20 + \tfrac{2}{3}d)$$

$$90{,}000 + 12{,}000d + 500d^2 + \frac{20}{d}d^3 = 10{,}000d^2 + \frac{1000}{d}d^3$$

Rearranging:

$$\frac{980}{3}d^3 + 9500d^2 - 12{,}000d - 90{,}000 = 0$$

$$0.98d^3 + 28.5d^2 - 36d - 270 = 0$$

By trial and error, $d \approx 3.5$ ft.

9.15 c.

Test No.	Normal Force, N	Shear Force, N	$\phi = \tan^{-1}\left(\dfrac{\text{Shear Force}}{\text{Normal Force}}\right)$, deg
1	271	120.6	24
2	406.25	170.64	22.8
3	474	204.1	23.3
4	541.65	244.3	24.3

Av. $\phi = 23.6°$

9.16 c.

$$\sigma_3' = 140 \text{ kN/m}^2$$
$$\sigma_1' = \sigma_3' + \Delta\sigma_{d(f)} = 140 + 263.5 = 403.5 \text{ kN/m}^2$$
$$\sigma_1' = \sigma_3' \tan^2\left(45 + \frac{\phi}{2}\right)$$

$$403.5 = 140\tan^2\left(45 + \frac{\phi}{2}\right)$$
$$\left(45 + \frac{\phi}{2}\right) = \tan^{-1}\sqrt{\frac{403.5}{140}} = 59.5°$$
$$\frac{\phi}{2} = 14.5°; \phi = 29°$$

9.17 c. Consider a 1 ft length of the slope. Weight of wedge,

$$W = \tfrac{1}{2}\gamma H^2 (\cot 15° - \cot 30°)$$

or

$W = \tfrac{1}{2}(108)(30)^2 (\cot 15° - \cot 30°) = 97{,}198$ lb
$N_r = N_a = W \cos 15° = (97{,}198)(\cos 15°) = 93{,}886$ lb
$T_a = W \sin 15° = (97{,}198)(\sin 15°) = 25{,}157$ lb

The maximum resistance that can be mobilized along AC is

$$T_r = c(\overline{AC}) + N_r \tan\phi = (500)\left(\frac{30}{\sin 15°}\right) + (93{,}886)(\tan 20°)$$
$$= 57{,}957 + 34{,}172 = 92{,}129 \text{ lb}$$

$$\text{Factor of safety} = \frac{T_r}{T_a} = \frac{92{,}129}{25{,}157} = 3.66$$

9.18 c. About O,

Driving moment for sliding = (32.4 kip)(16.1 ft) = 521.64 kip-ft

and

$$\text{Resisting moment} = r^2 c_u \theta = (36 \text{ ft})^2 (0.8 \text{ kip/ft}^2)\left(\frac{\pi}{180} \times 67\right) = 1212.4 \text{ kip-ft}$$

$$\text{The factor of safety} = \frac{\text{resisting moment}}{\text{driving moment}} = \frac{1212.4}{521.64} = 2.32$$

9.19 **d.** Refer to Exhibit 9.19a.

$$H = 1 \text{ ft} + 15 \text{ ft} + 8 \tan 10° = 17.41 \text{ ft}$$

Consider a 1 ft length of the retaining wall, as shown in Exhibit 9.19(a). The Rankine active force on $\overline{BC} = \frac{1}{2}\gamma H^2 K_a$. From Table 9.6 of the text for $\alpha = 10°$ and $\phi = 36°$, $K_a = 0.27$.

$$P_a = \frac{1}{2}(110)(17.41)^2(0.27) = 4501 \text{ lb/ft}$$

Overturning moment:

$$M_{A(O)} = (P_a)(\cos 10°)\left(\frac{H}{3}\right) = (4501)(\cos 10°)\left(\frac{17.41}{3}\right) = 25,724 \text{ ft-lb/ft}$$

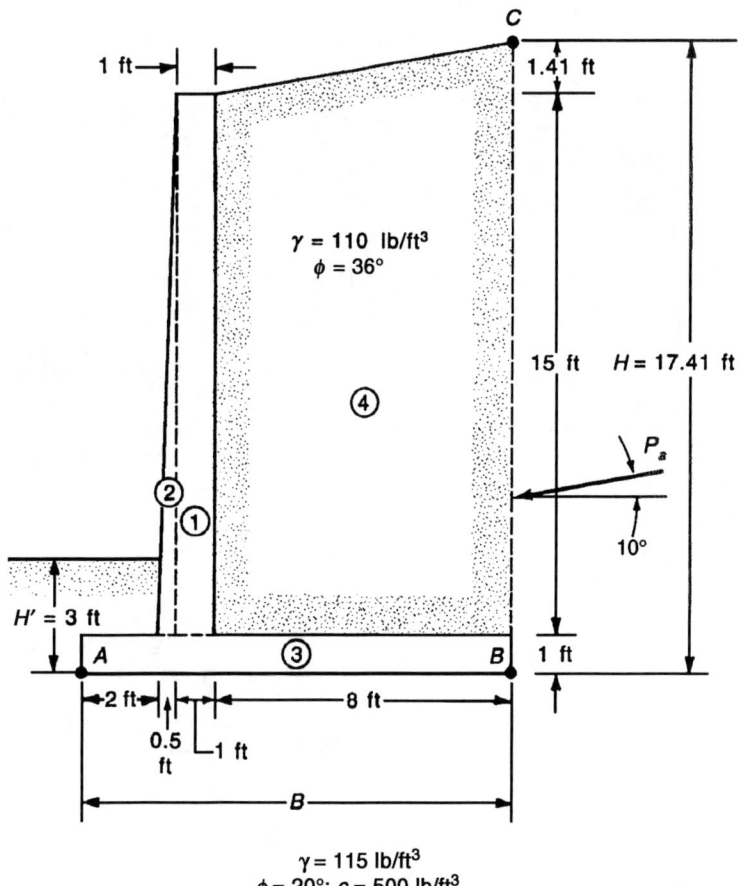

Exhibit 9.19a

Now the following table can be prepared:

Factor of safety against overturning

Section	Weight, lb/ft	Moment Arm from A, ft	Moment About A, ft-lb/ft
1	$(1)(15)(150) = 2{,}250$	3	6,750
2	$½(0.5)(15)(150) = 562.5$	2.33	1,310.6
3	$(11.5)(1)(150) = 1{,}725$	5.75	9,918.8
4	$½(15+16.41)(8)(110) = 13{,}820.4$	~7.5	103,653
	$P_a = 4501 \sin 10 = 781.6$	11.5	8,988.4
	$\Sigma V = 19{,}139.5$ lb/ft		$\Sigma M_{A(R)} = 130{,}620.8$

$$\frac{\Sigma M_{A(R)}}{\Sigma M_{A(O)}} = \frac{130{,}620.8}{25{,}724} = 5.08$$

9.20 **c.** Factor of safety against sliding

$$\frac{\Sigma V \tan \delta + B c_a + P_p}{P_a \cos \alpha}$$

P_p = Rankine passive pressure in front of toe = $½ K_p \gamma H'^2 + 2c\sqrt{K_p} H'$

$$K_p = \tan^2\left(45 + \frac{\phi}{2}\right) = \tan^2\left(45 + \frac{20}{2}\right) = 2.04$$

$\gamma = 115$ lb/ft^3; $c = 500$ lb/ft^2; $H' = 3$ ft

So

$P_p = ½(2.04)(115)(3)^2 + (2)(500)(\sqrt{2.04})(3) = 1055.7 + 4284.9 = 5340.6$ lb/ft

Also, V = weight/ft is 19,139.5 lb/ft (can be found from the table given in the solution of Problem 9.19).

$$\text{Factor of safety} = \frac{19{,}139.5 \tan 15° + (11.5)(300) + 5340.6}{4501 \cos 10°} = 3.14$$

REFERENCES

1. Boussinesq, J. *Application des Potentials à L'Etude de L'Equilibre et du Mouvement des Solides Elastiques.* Gauthier-Villars, Paris, 1883.
2. DeBeer, E. E. Experimental determination of the shape factors and bearing capacity factors of sand. *Geotechnique*, Vol. 20, No. 4, 1970, pp. 387–411.
3. Hansen, J. B. A revised and extended formula for bearing capacity. Danish Geotechnical Institute, *Bulletin 28*, Copenhagen, 1970.

4. Meyerhof, G. G. Penetration tests and bearing capacity of cohesionless soils. *Journal of the Soil Mechanics and Foundations Division*, American Society of Civil Engineers, Vol. 82, No. SM1, 1956, pp. 1–19.
5. Meyerhof, G. G. Bearing capacity and settlement of pile foundations. *Journal of the Geotechnical Engineering Division*, American Society of Civil Engineers, Vol. 102, No. GT3, 1976, pp. 197–228.
6. Terzaghi, K. *Theoretical Soil Mechanics*. Wiley, New York, 1943.
7. Vesic, A. S. Analysis of ultimate loads of shallow foundations. *Journal of the Soil Mechanics and Foundations Division*, American Society of Civil Engineers, Vol. 99, No. SM1, 1973, pp. 45–73.

RECOMMENDATIONS FOR FURTHER STUDY

1. Das, B. M. *Principles of Geotechnical Engineering*, 5th ed. Brooks/Cole, Pacific Grove, CA, 2002.
2. Das, B. M. *Principles of Foundation Engineering*, 5th ed. Brooks/Cole, Pacific Grove, CA, 2004.

CHAPTER 10

Transportation Engineering

PROBLEMS

The reader is encouraged to consult the "Recommended References" at the end of this chapter for additional practice problems. In this regard, the reader may find the example problems in the Highway Capacity Manual and the AASHTO, Asphalt Institute, and Portland Cement Association Pavement Design Manuals particularly valuable. For a more diverse selection of example problems, the reader can consult any standard transportation engineering textbook, such as Garber and Hoel or Roess et al.

10.1 Notes for an earthwork cross section are as follows

$$\frac{\text{C 4.0} \quad \text{C 4.2} \quad \text{C 3.3} \quad \text{C 3.0} \quad \text{C 2.5}}{36.0 \quad\quad 24.0 \quad\quad 0 \quad\quad 8.0 \quad\quad 30.0}$$

The roadbed is 40 ft wide. Cut slopes are at 1 ft vertical to 4 ft horizontal. The area of the cross section is approximately:
a. 160 ft^2
b. 180 ft^2
c. 200 ft^2
d. 240 ft^2

10.2 The table below gives earthwork cross-sectional areas for three successive location on a roadway:

Station	Earthwork Cross Section, m^2	
	Cut	Fill
8+30	0.0	28.0
8+70	8.0	10.0
9+00	22.0	0.0

161

If material shrinks by 15 percent (that is, volume of a unit mass is 15 percent less in a compacted fill than in the material's existing state), the net waste or borrow for this job is approximately:
a. Waste 560 m^3 (volume material occupies in existing state)
b. Waste 11 m^3 (volume material occupies in existing state)
c. Borrow 390 m^3 (volume material occupies in compacted fill)
d. Borrow 300 m^3 (volume material occupies in compacted fill)

10.3 In the mass diagram shown in Exhibit 10.3, the limit of economic haul is 2000 ft. Points for which the LEH spans the various loops are shown in the diagram. For the job shown, the total amount of borrow is approximately:
a. 4000 yd^3
b. 6000 yd^3
c. 10,000 yd^3
d. 12,000 yd^3

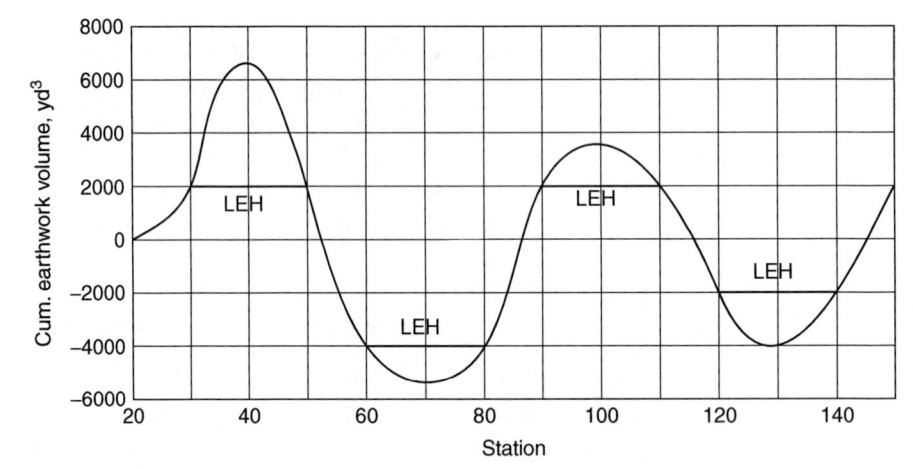

Exhibit 10.3

10.4 An existing two-lane highway between points A and B is to be converted to a four-lane divided freeway. Annual costs and benefits for four alternative improvement plans may be summarized as shown in Exhibit 10.4.

Exhibit 10.4 Summary of annual costs and benefits

	Existing	Plan 1	Plan 2	Plan 3	Plan 4
Costs:					
Equivalent annual highway costs ($)	30,000	410,900	561,300	702,000	1,520,000
Benefits:					
Time saving ($)		470,900	788,400	1,259,000	1,935,000
Accident saving ($)		182,200	191,000	193,700	210,000
Incremental operating costs ($)				−269,400*	−500,000
Annual benefits over existing highway ($)		653,100	979,400	1,183,300	1,645,000

* Increased user operating cost is a "disbenefit."

The preferred alternative is
a. Plan 1
b. Plan 2
c. Plan 3
d. Plan 4

10.5 The following questions refer to the horizontal circular curve shown in Exhibit 10.5. In answering the questions, the following specifications and data should be used: Design speed = 90 km/h; number of lanes = 2; maximum superelvation = 0.10; maximum side friction = 0.12.

    a. Based on the design speed of 90 km/h, the minimum radius of curvature is most nearly:
        a. 315 m
        b. 290 m
        c. 280 m
        d. 230 m

    b. If a 435 m curve is used, the sight distance would be most nearly:
        a. 175 m
        b. 180 m
        c. 190 m
        d. 192 m

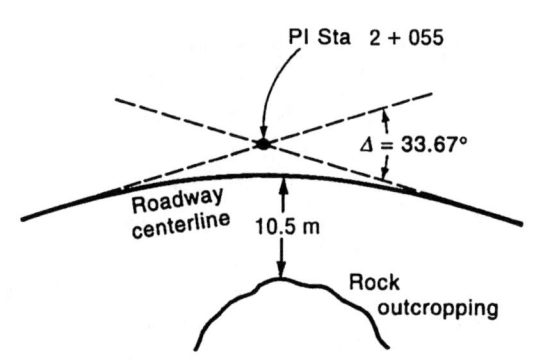

Exhibit 10.5 Horizontal roadway plan

(PI Sta 2 + 055; Δ = 33.67°; Roadway centerline; 10.5 m; Rock outcropping)

10.6 The centerline of an aqueduct was originally laid out as a reversed curve as indicated by the existing centerline in Exhibit 10.6. A construction project in the vicinity requires the aqueduct to be realigned. The new alignment will connect the existing tangents with a 360 m radius curve. This realignment will move the curved part of the centerline back away from the proposed project, and it will also replace the reversed curve that now exists.

    a. The central angle of the new curve is approximately:
        a. 30°
        b. 38°
        c. 45°
        d. 60°

Exhibit 10.6 Centerline plan

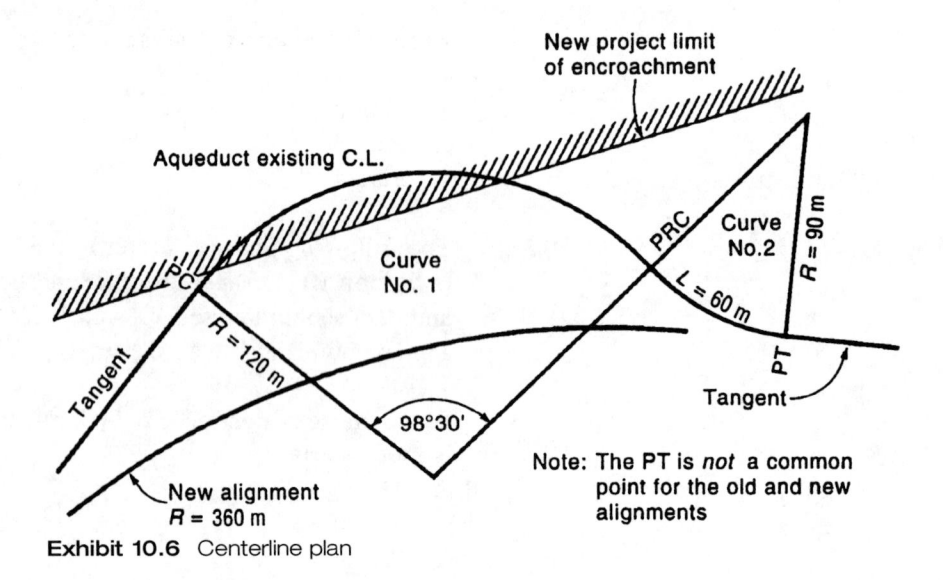

b. The distance between the PC of the new curve and the PC of the old curve is approximately:
  a. 250 m
  b. 210 m
  c. 190 m
  d. 120 m

c. The distance between the old PT and the new PT is approximately:
  a. 20 m
  b. 40 m
  c. 60 m
  d. 80 m

10.7 The profile shown in Exhibit 10.7 represents the preliminary design of a vertical curve that will pass underneath an existing freeway overpass. The minimum vertical clearance required between the roadway surface and the underside of the overpass is 4.9 m.

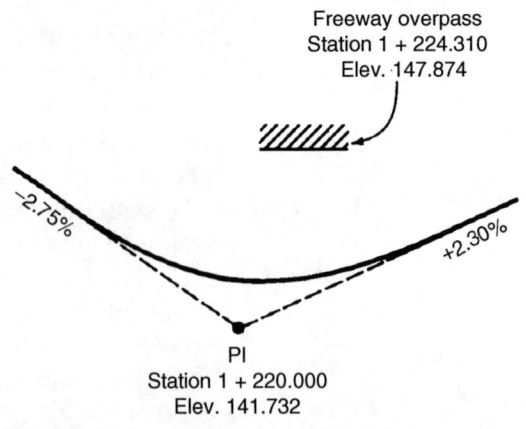

Freeway overpass
Station 1 + 224.310
Elev. 147.874

−2.75%   +2.30%

PI
Station 1 + 220.000
Elev. 141.732

Exhibit 10.7

a. Based on a 90 km/h design speed, the minimum length of the vertical curve is approximately:
   a. 180 m
   b. 192 m
   c. 205 m
   d. 250 m

b. Based on the specified minimum vertical clearance, the maximum length of the vertical curve is approximately:
   a. 128 m
   b. 178 m
   c. 198 m
   d. 278 m

c. For a 195 m vertical curve, the station and elevation of the low point of the curve are most nearly:
   a. Station = 1 + 220, Elevation = 143.0 m
   b. Station = 1 + 229, Elevation = 142.9 m
   c. Station = 1 + 075, Elevation = 144.4 m
   d. Station = 1 + 229, Elevation = 143.0 m

10.8 A driver approaches a two-way stop controlled intersection. The traffic on the main cross street has the right-of-way and an average gap of 6.5 s/veh. Assume that the driver can accelerate the vehicle such that $dv/dt = (3.6 - 0.06v)$ ft/s^2, the driver's perception time is 1.0 s, and the car is 20 ft long. The widest street the driver can safely clear is approximately:
   a. 48 ft
   b. 36 ft
   c. 33 ft
   d. 29 ft

10.9 A new private road is being planned to provide access to a proposed major ski resort area. In the preliminary design, a large outcropping of rock is located 30 ft from the centerline of a curved section of the proposed road. The grade on this curved section is 7 percent and the terrain and geologic conditions suggest that a curve radius of only 360 ft is economically practical. A superelevation rate of 10 percent will be used. A reasonable speed limit for this section of the road would be:
   a. 20 mph
   b. 25 mph
   c. 30 mph
   d. 35 mph

10.10 Exhibit 10.10 shows the basic profile for a proposed two-lane rural highway. The following data and specifications have been adopted as the basis for the preliminary design: ADT = 6000 vehicles; percent trucks = 10; entrance speed (at beginning of 2 percent grade) = 90 km/h; design vehicle = 120 kg/kW truck. If the AASHTO Green Book's 15 km/h speed-reduction criterion is used to determine the critical length of grades, the appropriate length of a truck-climbing lane for this section (including tapers) is about:
a. 420 m
b. 690 m
c. 720 m
d. 960 m

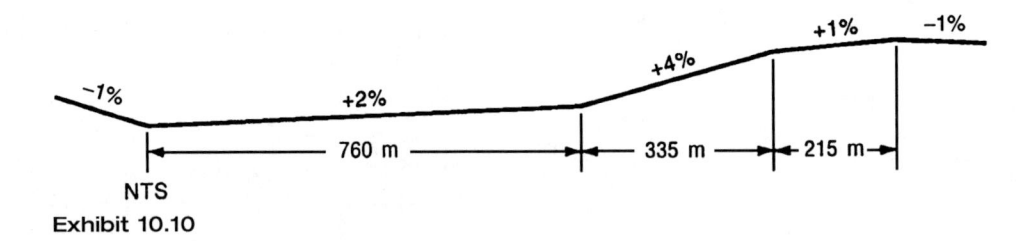

Exhibit 10.10

10.11 Results of a spot speed study are as summarized in the table below.

Speed Group (mph)	Number of Vehicles
<15	0
15–20	1
20–25	2
25–30	4
30–35	17
35–40	45
40–45	47
45–50	32
50–55	12
55–60	7
60–65	3
>65	0

In the absence of other information, an appropriate speed limit for this road would probably be:
a. 40 mph
b. 45 mph
c. 50 mph
d. 55 mph

10.12 An urban freeway has four lanes in the direction of travel being analyzed. It has a measured free-flow speed of 65 mph. The heavy vehicle factor is estimated to be 0.92 and the peak hour factor to be 0.90. The maximum hourly volume is expected to be 7500 veh/h. The level of service is
a. B
b. C
c. D
d. E

10.13 A rural freeway has two lanes in each direction. The free-flow speed is 110 km/h. The traffic stream consists of 10 percent trucks and buses and 5 percent recreational vehicles. The peak hour factor is 0.85, and the driver population factor is estimated to be 0.90. The maximum hourly volume that can be accommodated at level of service C on a 3.5 percent grade 2 km long is approximately:
a. 1720 vehicles/h
b. 2100 vehicles/h
c. 2550 vehicles/h
d. 3400 vehicles/h

10.14 A suburban highway consists of four lanes without a median. On the average, signalized intersections are more than 3.0 km apart on this roadway. Lane widths are 3.6 m, lateral clearance is 3.0 m, and there are approximately 12 access points per km. The speed limit is 90 km/h. The peak hour factor is 0.90. The heavy vehicle factor is estimated to be 0.92. If the hourly volume is 1700 veh/h, the level of service is:
a. A
b. B
c. C
d. D

10.15 Exhibit 10.15 shows a freeway weaving section and the weaving diagram with peak 15-minute flow rates in passenger cars per hour. Free-flow speed for the section is 65 mph.

Analysis of this weaving section has resulted in the following preliminary values: for unconstrained conditions, weaving speed = 51.1 mph and nonweaving speed = 60.2 mph. For constrained conditions, weaving speed = 42.7 mph, nonweaving speed = 65.1 mph, and N_w = 1.0. The estimated density in the weaving section is most nearly:
a. 23.8 pc/mi/ln
b. 24.7 pc/mi/ln
c. 27.8 pc/mi/ln
d. 32.2 pc/mi/ln

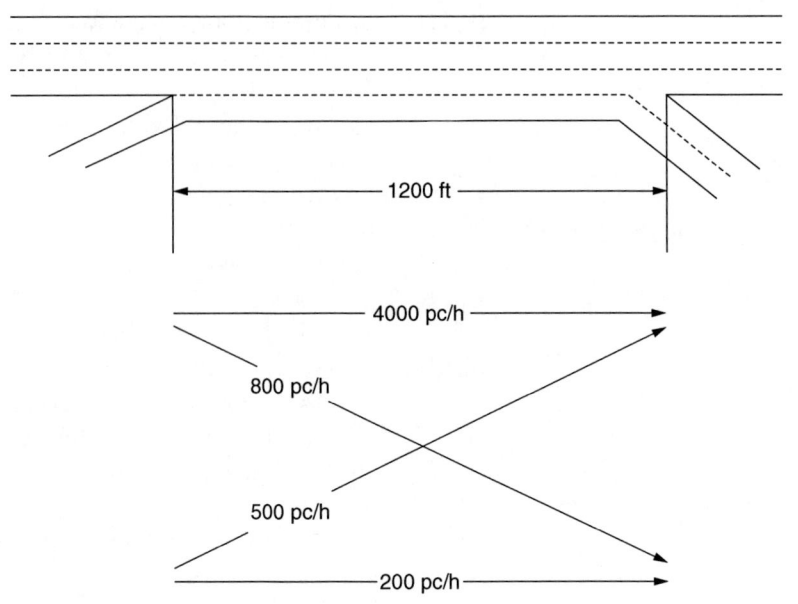

Exhibit 10.15

10.16 Exhibit 10.16 shows the lane configuration and peak 15-minute flow rates in pc/h for an isolated off-ramp. Free-flow speed is 100 km/h for the freeway and 60 km/h for the ramp.

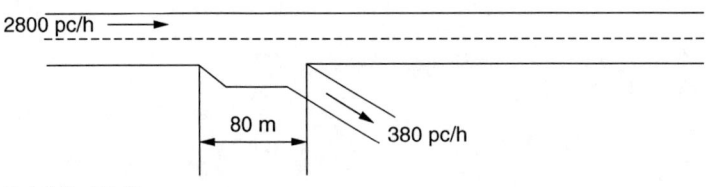

Exhibit 10.16

The level of service is
a. B
b. C
c. D
d. E

10.17 A Class I two-lane highway has a free-flow speed of 94.0 km/h. The highway is located in rolling terrain and has 60 percent no-passing zones. The directional split is 70/30. Adjusted two-way flow rates have been calculated and have been found to be 800 pc/h for purposes of calculating average travel speed and 1000 pc/h for purposes of calculating percent time spent following. The percent time spent following is most nearly:
a. 62.8
b. 63.8
c. 69.8
d. 71.8

10.18 A protected left-turn movement at an intersection in a suburban residential area utilizes a single-lane left-turn pocket. The turning lane is 11 feet wide and has a grade of +2 percent as it approaches the intersection. Traffic using the turn pocket includes 4 percent heavy vehicles. The saturation flow rate for the protected left turn is most nearly:
a. 1496 veh/h
b. 1662 veh/h
c. 1787 veh/h
d. 1800 veh/h

10.19 In an analysis of a two-way stop controlled intersection, one of the minor movements is found to have a flow rate of 30 pc/h and a capacity of 40 pc/h. If a 15-minute analysis period is used, the 95th percentile queue length for this movement is most nearly:
a. 3 vehicles
b. 4 vehicles
c. 10 vehicles
d. 20 vehicles

10.20 An intersection in an outlying business district has a 90-second cycle. Effective green time for pedestrians crossing one of the streets is 20 seconds. The level of service of this pedestrian movement is
a. B
b. C
c. D
d. E

SOLUTIONS

10.1 **b.** The cross section may be divided into triangles and trapezoids, as shown in Exhibit 10.1.

$$A = [(4.0 + 4.2)/2](12.0) + [(4.2 + 3.3)/2](24.0)$$
$$+ [(3.3 + 3.0)/2](8.0) + [(3.0 + 2.5)/2](22.0)$$
$$- [(4.0)(16.0)]/2 - [(2.5)(10.0)]/2$$
$$= 49.2 + 90.0 + 25.2 + 60.5 - 32.0 - 12.5 = 180.4 \text{ ft}^2$$

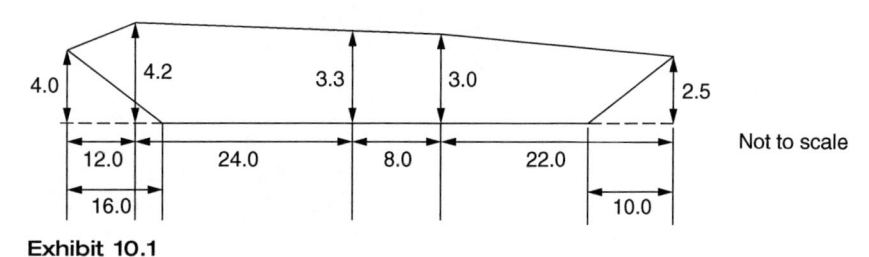

Exhibit 10.1

10.2 **c.** Between stations 8+30 and 8+70, the fill volume is calculated by the average end area method and the cut volume as a pyramid:

$$V_f = [(28 + 10)(40)]/(2) = 760.0 \text{ m}^3$$
$$V_c = [(8)(40)]/(3) = 106.7 \text{ m}^3$$

Between stations 8+70 and 9+00, cut is calculated by the average end area method and fill as a pyramid:

$$V_f = [(10)(30)]/(3) = 100.0 \text{ m}^3$$
$$V_c = [(8 + 22)(30)]/(2) = 450.0 \text{ m}^3$$
$$\text{Total cut} = 106.7 + 450.0 = 556.7 \text{ m}^3$$
$$\text{Total fill} = 760.0 + 100.0 = 860.0 \text{ m}^3$$

Accounting for shrinkage, the cut will provide $(556.7)(1.00 - 0.15) = 473.2 \text{ m}^3$ of fill.

Therefore, material will have to be borrowed. Borrow = $860.0 - 473.2 = 386.8 \text{ m}^3$ of compacted fill.

10.3 c. Portions of the mass diagram that have a positive slope indicate a surplus of cut over fill, and those with a negative slope indicate that more fill is required than can be supplied by the cut (if any) in that section. The longest distance material will be hauled is the limit of economic haul, in this case 2000 ft. Thus, any material represented by a negatively sloped portion of the mass diagram that is outside the portions of the loops cut off by the limit of economic haul lines must be borrowed. In the mass diagram in question, there are two such areas, as indicated in Exhibit 10.3a.

These total $6000 + 4000 = 10,000 \text{ yd}^3$.

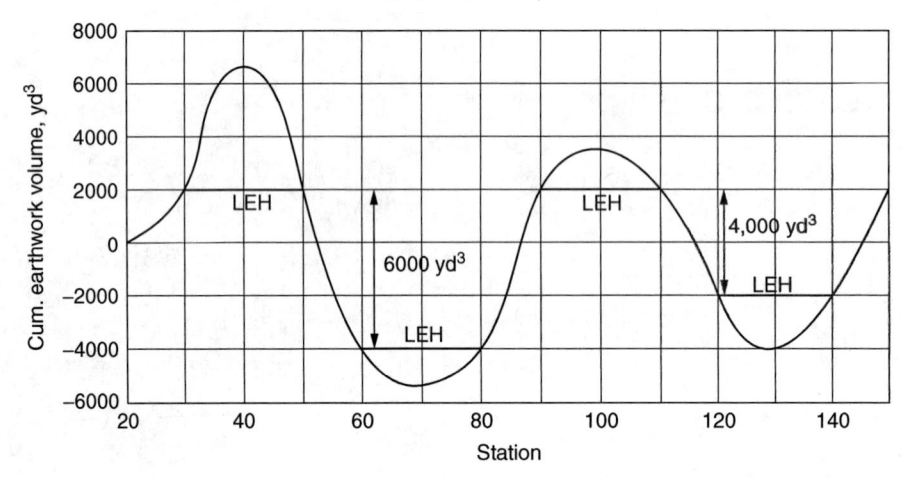

Exhibit 10.3a

10.4 c. Benefit/cost ratios:
Plan 1 versus Existing Highway
B/C = 653,100/(410,900 − 30,000) = 1.71 > 1.0. B/C > 1.0.
Therefore, Plan 1 is preferred over the existing highway.
Plan 2 versus Plan 1
B/C = (979,400 − 653,100)/(561,300 − 410,900) = 2.17. B/C > 1.0.
Therefore, Plan 2 is preferred over Plan 1.
Plan 3 versus Plan 2
B/C = (1,183,300 − 979,400)/(702,000) − 561,300 = 1.45. B/C > 1.0.
Therefore, Plan 3 is preferred over Plan 2.

Plan 3 versus Plan 4
B/C = (1,645,000 − 1,183,300)/(1,520,000 − 702,000) = 0.56.
B/C < 1.0.
Therefore, Plan 3 is preferred over Plan 4.

Conclusion: Plan 3 is the preferred plan.

10.5

a. b. From Eq. (10.21a) in *Civil Engineering PE License Review*,

$$R = V^2/127\,(e + f_s) = (90)^2/[127(0.10 + 0.12)] = 289.9 \text{ m}$$

b. d. From Exhibit 3-57 in AASHTO (2001), $S \approx 200$ m. Try 192 m:

$$M = 435\left[1 - \cos\left(\frac{28.96 \times 192}{435}\right)\right] = 10.55 \text{ m} \approx 10.5 \text{ m}$$

10.6

a. d. From Exhibit 10.6a of the text, Δ_3, which is also the central angle of the new curve, is $98.5° − \Delta_2$. Solving Eq. (10.90) for Δ, Δ_2 is given by

$$\Delta_2 = \frac{L_2(360°)}{2\pi R_2} = \frac{60(360°)}{2\pi(90)} = 38.2°$$

$$\Delta_3 = 98.5° − 38.2° = 60.3°$$

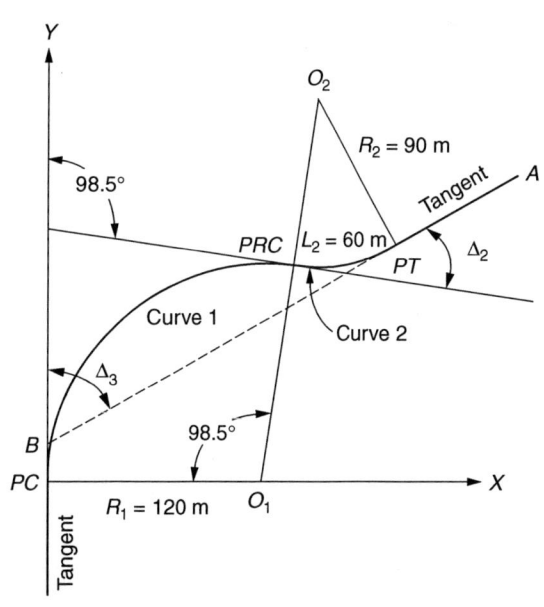

Exhibit 10.6a

b. c. From Exhibit 10.6b, the distance from the old *PC* to the new *PC* is
$T_3 - [T_1 - (d_1 + d_2)]$.

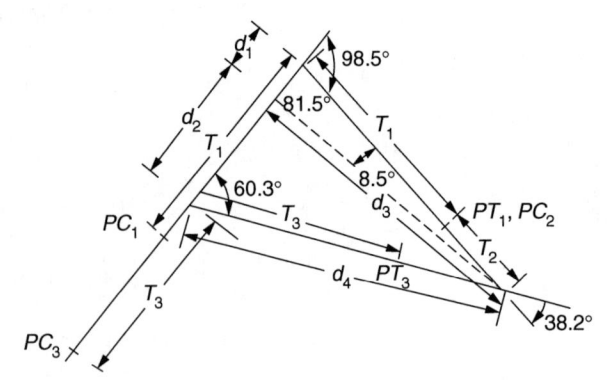

Exhibit 10.6b

$$d_1 = (T_1 + T_2) \sin 8.5°$$
$$d_3 = (T_1 + T_2) \cos 8.5°$$
$$d_2 = d_3 \cot 60.3°$$
$$T_1 = (120 \text{ m}) \tan (98.5°/2) = 139.3 \text{ m}$$
$$T_2 = (90 \text{ m}) \tan (38.2°/2) = 31.2 \text{ m}$$
$$T_3 = (360 \text{ m}) \tan (60.3°/2) = 209.1 \text{ m}$$
$$d_1 = (139.3 + 31.2) \sin 8.5° = 25.2 \text{ m}$$
$$d_3 = (139.3 + 31.2) \cos 8.5° = 168.6 \text{ m}$$
$$d_2 = 168.6 \cot 60.3° = 96.2 \text{ m}$$

The distance from the old *PC* to the new *PC* is

$$209.1 \text{ m} - [139.3 \text{ m} - (25.2 \text{ m} + 96.2 \text{ m})] = 191.2 \text{ m}$$

c. a. The distance from the old PT to the new PT is $d_4 + T_2 - T_3$,

$$d_4 = d_3/\sin 60.3° = (168.6 \text{ m})/\sin 60.3° = 194.1 \text{ m}$$
$$194.1 + 31.2 - 209.1 = 16.2 \text{ m}$$

10.7

a. b. Exhibit 3-75 of the AASHTO Green Book (2004, p. 277) suggests a design *K* value of 38 m per percent of algebraic difference in grade for sag vertical curves with a 90 km/h design speed. Equation (10.22) can be used to compute the minimum vertical curve length.

$$L = KA = 38[2.30\% - (-2.75\%)] = 191.9 \text{ m}$$

b. c. Based on the required vertical clearance of 4.9 m, the highest allowable curve elevation is 147.874 m − 4.9 m, or 142.974 m at STA 1 + 224.310. The maximum length of vertical curve with an elevation not exceeding 142.974 m at STA 1 + 224.310 can be determined as follows:

The horizontal distance from the PVI to the critical point is

(1 + 224.310) − (1 + 220.00) = 0 + 004.310 STA = 4.310 m

Extend the -2.75 percent grade from the PVI to STA 1 + 224.310. At this point, its elevation is

$El = E_{PVI} + (-0.0275)(4.310) = 141.613$ m

The difference between the roadway surface elevation and this elevation, 142.974 m − 141.613 m = 1.361 m, is the offset of the vertical curve. From Eq. (10.122) in *Civil Engineering PE License Review*, this is $(A/2L)x^2$. The horizontal distance from the PVC to the PVI is $L/2$, so $x = L/2 + 4.310$ m. Equating the formula for the offset to 1.361 m gives

$1.361 = (A/2L)(L/2 + 4.310)^2 = (0.0505/2L)(L/2 + 4.310)^2$

Expanding and rearranging terms leads to

$0.012625L^2 - 2.504345L + 0.93809395 = 0$

The solution of this quadratic equation is

$L = \{2.504345 \pm [(2.504345)2 - 4(0.012625)(0.93809395)]^{1/2}\}/2(0.012625)$

The large root of the quadratic, $L = 197.989$ m ≈ 198 m, is the maximum length of the vertical curve.

Also, if the value of K is specified, it is possible to solve directly for L using $L = KA$. Exhibits 3-72, 3-73, and 3-75 [pages 272 and 277 of the Green Book (2004)] give recommended design values for K.

Note that the recommended design values for K given in Exhibits 3-72, 3-73, and 3-75 are based on the case in which $S < L$. Values of L based on $L = KA$ will be conservative if $S > L$.

Note: this problem can be simplified by use of a formula given by Banks (2002, p. 75).

c.   **b.** Use Eq. (10.126) in *Civil Engineering PE License Review* to compute the distance from the PVC to the low point.

$X_{LP} = (-g_1 L)/(g_2 - g_1) = [-(-2.75)(0.195)]/[2.30 - (-2.75)] = 0.106188$

Compute the STA of the low point.

Low Point = PVC + X_{LP} = [(1 + 220.000 − (0 + 097.500)] + (0 + 106.188)
= 1 + 228.688

Use Eq. (10.123) in *Civil Engineering PE License Review* to compute the elevation of the low point.

$E_{LP} = E_{PVC} + g_1 X + [g_2 - g_1)/2L]X^2 = 144.413 + (-0.0275)(106.186)$
$+ \{[0.0230 - (-0.0275)]/(2)(195)\}(160.188)^2 = 142.934$ m

10.8   **d.** The distance-time relationship for nonuniform acceleration is given in Eq. (10.8) in *Civil Engineering PE License Review*.

$d = (\alpha t/\beta) - (\alpha/\beta^2)(1 - e^{-\beta t}) + (v_0/\beta)(1 - e^{-\beta t})$

From the problem statement, $\alpha = 3.6$, $\beta = 0.06$, $t = 6.5$ s $- 1$ s PIEV time $= 5.5$ s, and $v_0 = 0$. Therefore:

$$d = [(3.6)(5.5)/0.06] - [3.6/(0.06)^2]\left(1 - e^{-0.06(5.5)}\right) + 0 = 48.9 \text{ ft}$$

Subtracting the length of the vehicle, the widest street that can be cleared is $48.9 - 20 = 28.9$ ft.

10.9 d. From Eq. 3-38 on p. 227 of the AASHTO Green Book (2004), the relationship between the middle ordinate M, radius R, and sight distance S is

$$M = R[1 - \cos(28.65 S/R)]$$

Therefore,

$$\begin{aligned} S &= (R/28.65)\cos^{-1}[(R-M)/R]) \\ &= (360/28.65)\cos^{-1}[(360-30)/360] \\ &= 296.0 \text{ ft} \end{aligned}$$

To solve this problem, it is necessary to determine whether the speed is controlled by sight distance or the radius. The speed as a function of sight distance is computed from Eq. (10.13b), in *Civil Engineering PE License Review*

$$d_s = 1.47Vt + V^2/30[(a/32.2) \pm G]$$

Assuming a PIEV time of 2.5 s and $a = 11.2$ ft/s^2 (as recommended by the Green Book, p. 113) gives

$$296.0 = 3.675\,V + V^2/30[(11.2/32.2) + 0.07]$$

or

$$0.08V^2 + 3.675V - 296.0 = 0$$

From the quadratic equation,

$$V = \{-3.675 \pm [(3.675)^2 - (4)(-296.0)(0.08)]^{0.5}/[(2)(0.08)] = 42.1 \text{ mph}$$

Equation (10.21b) in *Civil Engineering PE License Review* defines the relationship between speed and the radius

$$R = V^2/[15(f_s + e)] \text{ and } V = [15\,R(f_s + e)]^{0.5}$$

Assuming $e = 0.10$ and $V = 40$ mph, the Green Book (Exhibit 3.15, p. 147) gives $f_s = 0.16$. Solving for V,

$$V = [15R(0.16 + 0.10)]^{0.5} = [(15)(360)(0.16 + 0.10)]^{0.5} = 37.5 \text{ mph}$$

The safe speed for the 360 ft radius is less than that to satisfy the sight distance requirement. Therefore, an appropriate speed limit would be 35 mph.

10.10 b. From Exhibit 3-55, p. 234 of the AASHTO Green Book (2004), the truck will decelerate along the 2 percent grade over its 760 m length to about 84 km/h. Again from Exhibit 3-55, the speed of the truck at the end of the 4 percent grade is about 70 km/h. At the end of the 215 m long grade of 1 percent, the truck will have accelerated from 70 km/h (speed at end of 4 percent grade) to about 72 km/h (Exhibit 3-56).

The next step in the solution is to enter Exhibit 3-56 (p. 235 in the AASHTO Green Book) and move horizontally to the right to the next grade line, which is −1 percent. Move along this grade line vertically and to the right to its intersection with the 75 km/h line. The horizontal distance from the intersections of the −1 percent grade line with the 72 km/h and 75 km/h horizontal lines is 90 m. This represents the end of the required climbing lane. The climbing lane begins 220 m from the start of the 4 percent grade. It continues 115 m along the 4 percent grade, 215 m along the +1 percent grade, and 90 m along the −1 percent grade. So the total length of the climbing lane is 420 m.

Adding the recommended transitions (see p. 246, 2004 Green Book) on the front end (25:1 taper or 90 m) and on the back end (50:1 taper or 180 m) results in a total length, including tapers, of 690 m.

10.11 c. The Manual on Uniform Traffic Control Devices (FHWA, 2003) advises that speed limits should be within 5 mph of the 85th percentile speed (p. 2B-10). The 85th percentile speed may be calculated by calculating the cumulative speed distribution, expressed in percent, as shown in the table below.

Speed Group (mph)	Number of Vehicles	Cumulative Vehicles	Cumulative Percent
<15	0	0	0.0
15–20	1	1	0.6
20–25	2	3	1.8
25–30	4	7	4.1
30–35	17	24	14.1
35–40	45	69	40.6
40–45	47	116	68.2
45–50	32	148	87.1
50–55	12	160	94.1
55–60	7	167	98.2
60–65	3	170	100.0
>65	0	170	100.0

The 85th percentile speed is between 45 and 50 mph. Rounding to the nearest 5 mph increment, the speed limit should be 50 mph.

10.12 d. Since this is an urban freeway, assume $f_p = 1.00$.

$$v_p = 7500/[(4)(0.90)(0.92)(1.00)] = 2264 \text{ pc/h/ln}$$

From Exhibit 23.2 (TRB, 2000), for a free-flow speed of 65 mph, the maximum service flow rate for level of service D is 2090 pc/h/ln, and the maximum service flow rate for level of service E is 2350 pc/h/ln.

$$2090 < 2264 < 2350, \text{ so the level of service is E}$$

10.13 b. From Exhibit 23-2 in the Highway Capacity Manual (2000), the maximum service flow rate at level of service C for a freeway with a free-flow speed of 110 km/h is 1740 pc/h/ln. For the 3.5 percent grade, $E_T = 3.0$ and $E_R = 2.5$. The heavy vehicle factor is

$$f_{HV} = 1/[1 + P_T(E_T - 1) + P_R(E_R - 1)]$$
$$= 1/[1 + 0.10(3.0 - 1) + 0.05(2.5 - 1)] = 0.784$$

PHF = 0.85, $N = 2$, and $f_p = 0.90$. Therefore, the maximum hourly volume is $V = v_p \text{ PHF } N f_{HV} f_p = (1740)(0.85)(2)(0.784)(0.90) = 2087$ veh/h.

10.14 c. This roadway should be analyzed as a suburban multilane highway using the procedures in Chapter 21 of the HCM. Since the free-flow speed is not known, it must be estimated by FFS = BFFS $- f_{LW} - f_{LC} - f_M - f_A$. According to pp. 21–25 of the HCM, the base free-flow speed for this type of facility is about 8 km/h above the speed limit for a speed limit of 90 km/h. The adjustment for lane width is 0.0 (Exhibit 21.4), the adjustment for lateral clearance is 0.6 km/h (Exhibit 21.5), the adjustment for median type is 2.6 km/h, and the adjustment for access-point density is 8.0 km/h:

$$\text{FFS} = 98 - 0.0 - 0.6 - 2.6 - 8.0 = 86.8 \text{ km/h}$$

The 15-minute passenger car equivalent flow rate is $v_p = V/(\text{PHF } N f_{HV} f_p)$. Because this is a suburban facility, assume that $f_p = 1.00$.

$$v_p = 1700/[(0.90)(2)(0.92)(1.00)] = 1027 \text{ pc/h/ln}$$

From Fig. 10.16b of the text (Exhibit 21-3 of the HCM), the level of service of FFS = 86.8 km/h and $v_p = 1027$ pc/h/ln is C.

10.15 a. From the Exhibit 10.15, on-ramp vehicles must change lanes once to enter the freeway, but off-ramp vehicles may exit the freeway without changing lanes. Consequently, the section is a Type B weaving section. For Type B weaving sections, the maximum value of N_w

for unconstrained operation is 3.5. $N_w = 1.0 < 3.5$, so operation is unconstrained. The average speed is

$$S = v/[(v_w/S_w) + (v_n/S_n)]$$

Total weaving flow is $800 + 500 = 1300$ pc/h, total nonweaving flow is $4000 + 200 = 4200$ pc/h, and total flow in the section is $1300 + 4200 = 5500$ pc/h. Consequently, $S = 5500[(1300/51.1) + (4200/60.2)] = 57.8$ mph.

The density is given by $D = (v/N)/S$. The number of lanes in the weaving section is four, so $D = (5500/4)/57.8 = 23.8$ pc/mi/ln.

10.16 b. The freeway has two lanes in the direction of travel. From Exhibit 25-12 of the HCM,

$$P_{FD} = 1.00.$$
$$v_{12} = v_R + (v_F - v_R)P_{FD} = 380 + (2800 - 380)1.00 = 2800 \text{ pc/h}$$

Maximum flow on the freeway occurs upstream of the ramp. Checking the capacity from Exhibit 25.14, $2800 < 4600$, so the section is under capacity. Also, $2800 < 4400$, so the flow entering the influence area is less than the maximum desirable. The density is given by

$$D_R = 2.642 + 0.0053v_{12} - 0.0183L_D = 2.642 + 0.0053(2800) - (0.0183)(80)$$
$$= 16.0 \text{ pc/km/ln}$$

From Exhibit 25-4, the upper limit of level of service B is 12 pc/km/ln, and the upper limit of level of service C is 17 pc/km/ln. $12 < 16.0 < 17$, so the level of service is C.

10.17 c. Base percent time spent following is given by BPTSF = $100[1 - \exp^{-0.000879(v_p)}]$. The adjusted flow rate to be used in calculating percent time spent following is 1000 pc/h. Therefore, BPTSF = $100[1 - \exp^{-0.000879(1000)}] = 58.5$. Percent time spent following is given by PTSF = BPTSF + $f_{d/np}$. For cases with a directional split of 70/30 and 60 percent no-passing zones, Exhibit 20-12 of the HCM gives values of $f_{d/np}$ of 13.3 for $v_p = 800$ and 7.4 for $v_p = 1400$. Interpolating to get a value for $v_p = 1000$, $f_{d/np} = 13.3 + [(1000 - 800)/(1400 - 800)] (7.4 - 13.3) = 11.3$. PTSF = $58.5 + 11.3 = 69.8$.

10.18 b. Saturation flow is given by

$$s = s_0 \, N f_w \, f_{HV} \, f_g \, f_p \, f_{bb} \, f_a \, f_{LU} \, f_{RT} \, f_{LT} \, f_{Rpb} \, f_{Lpb}$$

In the case, adjustment factors that apply are those for lane width, heavy vehicle presence, grade, and area type. In addition, $f_{LT} = 0.95$ for protected left turns. All other factors are 1.00; since this is a protected left turn, there will be no right-turn factor or interference with pedestrians, bicycles, buses, or parked vehicles; since there is only one lane, the lane utilization factor is 1.00. Of the factors that apply, $f_a = 1.00$ because the intersection is not located in a central business district. Other factors are as follows:

$$f_w = 1+[(W-12)/30] = 1+[(11-12)/30] = 0.967$$
$$f_{HV} = 100/[100+\%HV(E_T-1)] = 100/[4(2-1)] = 0.962$$
$$f_g = 1-(\%G/200) = 1-(2/200) = 0.99$$
$$s = (1900)(1)(0.967)(0.962)(0.99)(1)(1)(1)(1)(0.95)(1)(1) = 1662 \text{ veh/h}$$

10.19 a. The 95th percentile queue is given by Eq. (17-37) in the HCM as

$$Q_{95} \approx 900T\{v_x/c_{m,x}-1+[(v_x/c_{m,x}-1)^2+(3600/c_{m,x})(v_x/c_{m,x})/(150T)]^{0.5}\}(c_{m,x}/3600)$$
$$= (900)(0.25)\{30/40-1+[(30/40-1)^2+(3600/40)(30/40)/(150 \times 0.25)]^{0.5}$$
$$(40/3600)$$
$$= 2.78$$

10.20 b. Average pedestrian delay at a signalized intersection is given by Eq. (18-5) in the HCM as

$$d_p = [0.5(C-g)^2]/C = [0.5(90-20)^2]/90 = 27.2 \text{ s/pedestrian}$$

From Exhibit 18.9 of the HCM, the level of service is C.

REFERENCES

1. American Association of State Highway and Transportation Officials (AASHTO). *AASHTO Guide for Design of Pavement Structures*. Washington, DC, 1993.
2. American Association of State Highway and Transportation Officials (AASHTO). *Roadside Design Guide*. Washington, DC, 2002.
3. American Association of State Highway and Transportation Officials (AASHTO). *A Policy on Geometric Design of Highways and Streets*, Washington, DC, 2004.
4. Asphalt Institute. *Thickness Design: Asphalt Pavements for Highways and Streets*, Manual Series No. 1 (MS-1). Asphalt Institute, Lexington, KY, Feb. 1991.
5. Banks, J. H. *Introduction to Transportation Engineering*, 2nd ed. McGraw-Hill, New York, 2002.
6. Brinker, R. C., and Minnick, R. *The Surveying Handbook*, 2nd ed. Chapman and Hall, New York, 1995.
7. Daniel, W. W. *Applied Nonparametric Statistics*, Houghton Mifflin, Boston, 1978.
8. Drew, D. R. *Traffic Flow Theory and Control*. McGraw-Hill, New York, 1968.
9. Fambro, D. B., Chang, E. C. P., and Messer, C. J. *Effects of the Quality of Progression on Delay*. National Cooperative Highway Research Program (Report No. 339, TRB). Washington, DC, Sept. 1991.
10. Garber, N. J. and Hoel, L. A. *Traffic and Highway Engineering*, 2nd ed. PWS Publishing, Boston, MA, 1997.
11. Hickerson, T. F. *Route Location and Design*, 5th ed. McGraw-Hill, New York, 1964.
12. Institute of Transportation Engineers (ITE). *Parking Generation*, 2nd ed. Washington, DC, 1987.

13. Institute of Transportation Engineers (ITE). *1991 Membership Directory*. Washington, DC, 1991.
14. Institute of Transportation Engineers (ITE). *Trip Generation*, 7th ed. Washington, DC, 2003.
15. Institute of Transportation Engineers (ITE). *Traffic Engineering Handbook*, 5th ed. Washington, DC, 1999.
16. Ives, H. C. and Kissam, P. *Highway Curves*, 4th ed. John Wiley and Sons, New York, 1952.
17. Kavanagh, B. F. *Surveying Principles and Applications*, 4th ed. Prentice Hall, Englewood Cliffs, NJ, 1995.
18. Roess, R. P., Prassas, E. S. and McShane, W. R. *Traffic Engineering*, 3rd ed. Prentice Hall, Englewood Cliffs, NJ, 2004.
19. Meyer, M. D. and Miller, E. J. *Urban Transportation Planning: A Decision-Oriented Approach*, 2nd ed. McGraw-Hill, New York, 2001.
20. Portland Cement Association (PCA). *Thickness Design for Concrete Highway and Street Pavements*, PCA. Skokie, IL, 1984.
21. Robertson, H. D. *Manual of Transportation Engineering Studies*. Institute of Transportation Engineers, Prentice Hall, Englewood Cliffs, NJ, 2000.
22. Transportation Research Board (TRB). *Travel Estimation Techniques for Urban Planning*. (National Cooperative Highway Research Program Report No. 365). Washington, DC, 1998.
23. Transportation Research Board (TRB). *Highway Capacity Manual 2000*, Washington, DC, 2000.
24. U.S. Department of Transportation (USDOT), Federal Highway Administration, Federal Transit Administration. *An Introduction to Urban Travel Demand Forecasting: A Self-Instructional Text*. Washington, DC, 1977.
25. U.S. Department of Transportation (USDOT), Federal Highway Administration. *TRANSYT-7F: Traffic Network Study Tool (Version 7F). Self-Study Guide*. Washington, DC, July 1986.
26. U.S. Department of Transportation (USDOT), Federal Highway Administration. *Manual on Uniform Traffic Control Devices for Streets and Highways*. Washington, DC, 2003. (May be downloaded at *http://mutcd.fhwa.dot.gov/*.)
27. Wadsworth, H. M., Jr., Ed. *Handbook of Statistical Methods for Engineers and Scientists*. McGraw-Hill, New York, 1990.
28. Webster, F. V., and Cobbe, B. M. *Traffic Signals* (Road Research Technical Paper No. 39). Her Majesty's Stationery Office, London, 1958.

RECOMMENDED EXAM REFERENCES

The licensing examination is open book, and most states do not limit the number of books that the examinee can bring to the exam. However, the exam is very fast-paced, and there is generally not enough time to use books with which the examinee is not already thoroughly familiar. Therefore, is it extremely important to be selective in choosing reference materials. With this in mind, the author recommends the following basic references for the reader's consideration. Depending upon the reader's background, additional references should be selected as deemed necessary.

1. **Garber and Hoel,** *Traffic and Highway Engineering,* **1997.** This textbook provides an excellent overview of the major topics reviewed in this chapter, and it contains the key design charts and equations for many of the procedures, such as highway capacity procedures, AASHTO geometric and pavement design procedures, and the Asphalt Institute and Portland Cement Association pavement design methods. The textbook is particularly strong in its treatment of pavement design procedures. The treatment of the various topics includes numerous example problems with step-by-step solutions.

2. *AASHTO Policy on Geometric Design of Highways and Streets,* **2004.** This is the standard reference on highway design, and the reader should become familiar with the basic organization of this voluminous manual prior to the exam.

3. **Hickerson,** *Route Location and Design,* **1964.** This is the classic text on the topic. It provides extensive and detailed treatment of route surveying and related topics, including numerous example problems with step-by-step solutions.

4. **Brinker and Minnick,** *The Surveying Handbook,* **1995.** This is a comprehensive, up-to-date reference on surveying practice and includes material on the use of the metric system in surveying.

5. **Roess, Prassas, and McShane,** *Traffic Engineering,* **2004.** An excellent overview of the state of the art. The treatment of highway capacity analysis is especially insightful and well documented. Includes most of the key equations, design charts, and tables needed to apply the 2000 HCM procedures. It is extensively illustrated with example problems and solutions.

CHAPTER 11

Construction Engineering

PROBLEMS

11.1 A client requires a new facility quickly to replace one destroyed by a hurricane. The client is not certain exactly what it should include. What type of contract should be used?
   a. Unit price
   b. Lump sum
   c. Cost plus fixed fee
   d. Surety

11.2 Determine the hourly labor cost for a steelworker who earns $30 per hour and has the following benefits:

Fringes:	$3 per hour
Worker's Comp:	$12 per $100 of pay
PL and PD:	$5 per $100 of pay
FICA:	6.2%
Unemployment:	5%
Subsistence:	$40 per day

The steelworker works five 8-hour shifts per week.
   a. $33.00/hr
   b. $50.00/hr
   c. $46.45/hr
   d. $58.00/hr

11.3 A dump truck can haul 20 loose cubic yards of clay. Using typical shrinkage and swell values for this material, how many truck loads will be required to haul enough clay to provide 1000 compacted cubic yards of backfill?
   a. 82
   b. 40
   c. 63
   d. 50

181

11.4 Determine the volume of fill for a whole station (100') of cut with the end areas shown in Exhibit 11.4. Use the prismoidal method.

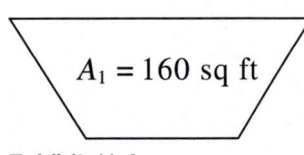

$A_1 = 160$ sq ft $A_m = 200$ sq ft $A_2 = 120$ sq ft

Exhibit 11.4

a. 592.6 cy
b. 480 cy
c. 960 cy
d. 666.7 cy

11.5 A roadbed is to be compacted to the specifications shown in the compaction curve in Exhibit 11.5.

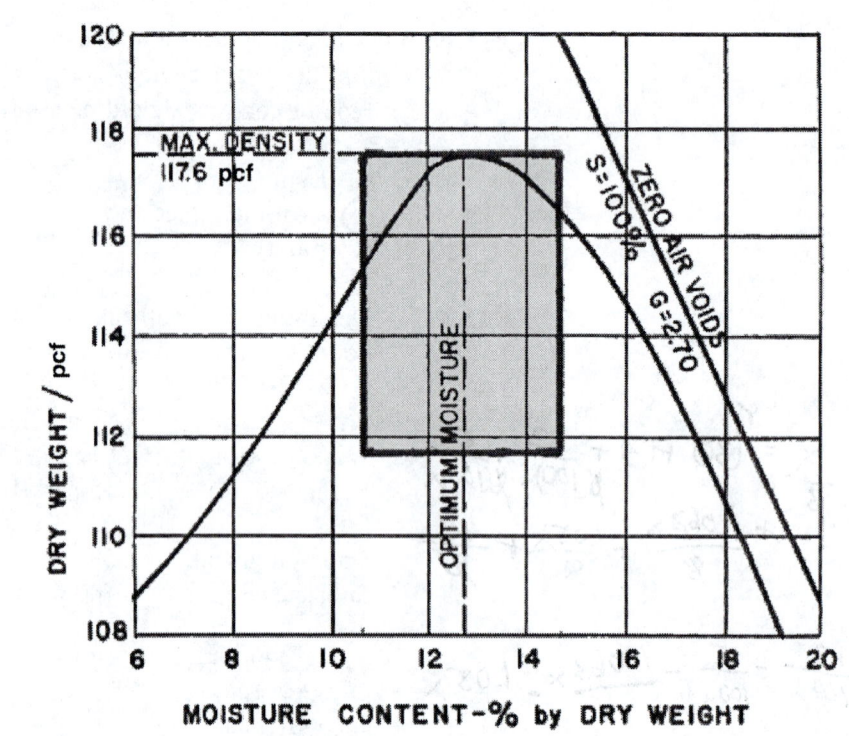

Exhibit 11.5

A sand cone test removed 1.4 lbs of compacted soil from a hole. The hole was measured to have a volume of 0.012 cf. The soil weighed 1.2 lbs after drying. What corrective action should be taken to meet specifications?

a. None
b. Add water and compact.
c. Compact more at current moisture content.
d. Dry soil and recompact.

11.6 A test strip shows that a steel-wheeled roller, operating at 3 mph, can compact a 0.5 ft. layer of material to proper density in four passes. The width of the drum is 8.0 ft. The roller operates 50 minutes per hour. The number of rollers required to keep up with a material delivery rate of 540 bank cubic yds/hr is (1 bank cubic yard = 0.83 ccy):
a. 1
b. 2
c. 3
d. 4

11.7 An excavator with a 3 cy bucket loads hard clay into dump trucks. Due to arrival of the trucks, the excavator works 45 minutes per hour. The clay has a swell of 35 percent. The production rate for the excavator is
a. 344 bcy/hr
b. 340 bcy/hr
c. 275 bcy/hr
d. 255 bcy/hr

11.8 Using the sum-of-years-digits method, what is the second year's depreciation on a wheeled loader that cost $100,000 with a salvage value of $12,000? Tires cost $8,000 for a set.
a. $10,666.67
b. $21,333.33
c. $23,466.67
d. $17,600

Problems 11.9–11.10
Use the CPM diagram and durations in Exhibit 11.9 to answer problems 11.9 and 11.10.

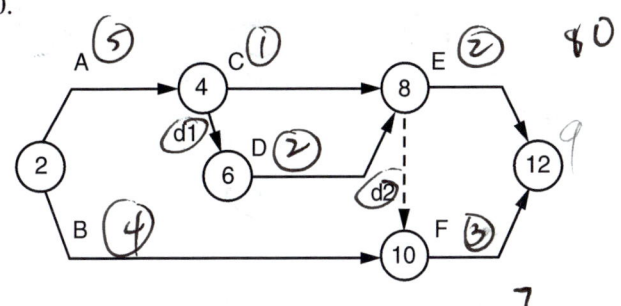

Activity	Duration
A	5 days
B	4 days
C	1 day
D	2 days
E	2 days
F	3 days

Exhibit 11.9

11.9 The activities on the critical path are
a. A, C, E
b. A, D, E
c. A, D, F
d. A, C, F

11.10 The duration of this project is
a. 8 days
b. 9 days
c. 10 days
d. 11 days

Problems 11.11–11.14

Use the following information to answer problems 11.11–11.14. The table shows the five critical activities and the optimistic, most likely, and pessimistic times for these activities. Use PERT procedures.

Critical Activity	min a	most likely b	max m
A	6	6	6
D	2	5	8
E	2	3	7
F	6	8	10
I	2	2	5
	18	24	36

$$\mu = \frac{1}{6}(18 + 4(24) + 36) = 25$$

$$\sigma = \frac{1}{6}(36 - 18) = 3$$

$$z = \frac{24 - 25}{3} = -\frac{1}{3}$$

11.11 The expected time of completion is:
a. 24 days
b. 25 days
c. 26 days
d. 27 days

11.12 The probability that the project will be completed by the end of day 26 is:
a. 25%
b. 50%
c. 75%
d. 100%

11.13 The probability of completion before the start of day 25 is
a. 26%
b. 75%
c. 50%
d. 23%

11.14 The completion day with at least 93 percent confidence is
a. 26
b. 27
c. 28
d. 29

SOLUTIONS

11.1 c. Cost plus fixed fee. Unit price and lump sum require a detailed design. Surety is a bond.

11.2 c. Wages: 5 days × 8 hrs/day = 40 hrs/week = regular pay

$$40 \text{ hrs} \times \$30.00 = \$1200$$
$$\text{Fringes: } \$3.00 \times 40 \text{ hrs} = \$120$$
$$\text{Worker's Comp + PD + PL} = \$12 + \$5 = \$17 \text{ per } \$100 \text{ base pay}$$
$$= \$17 \times \$1200/100 = \$204$$
$$\text{FICA} = 0.062 \times \$1200 = \$74.40$$
$$\text{Unemployment} = 0.05 \times \$1200 = \$60$$
$$\text{Subsistence} = \$40 \times 5 \text{ days} = \$200$$

Total = 1200 + 120 + 204 + 74 + 60 + 200 = $1858/40 hrs = $46.45 per hour

11.3 a. Typical shrinkage factor for clay is 0.80. This means that 1000 ccy/0.80 = 1250 bank cubic yards are required. The swell factor is 0.77. This means that 1250/0.77 = 1623.4 loose cubic yards must be hauled. This requires 1623.4/20 = 81.2, or 82 truck loads.

11.4 d. Volume = (1 cy/27 cf) 100 ft [160 sq. ft. + (4 × 200 sq ft) + 120 sq. ft.]/6 = 666.7 cy

$$\text{moisture content} = (112 \text{ g.} - 102 \text{ g.})/102 \text{ g.} \times 100\% = 9.8\%$$
$$\text{dry density} = 112.8 \text{ pcf}/1.098 = 102.7 \text{ pcf}$$

11.5 d. Wet density = 1.40/0.012 cf = 116.7 pcf

$$\text{Moisture content} = (1.4 - 1.2)/1.2 = 0.167 \times 100\% = 16.7\%$$
$$\text{Dry density} = 116.7/(1 + 0.167) = 100 \text{ pcf}$$

Specifications require dry density of greater than 111.8 pcf @ 10.4 – 14.4% moisture.

Soil is too wet and not dense enough. Therefore, soil must be dried until it is less than 14.4 percent moisture content and be recompacted to at least 111.8 pcf.

11.6 a. First calculate the roller production:

$$\text{ccy/hr} = \frac{16.3 \times 8 \text{ ft} \times 3 \text{ mph} \times 6 \text{ inches} \times 50 \text{ min}/60 \text{ min}}{4 \text{ passes}} = 489 \text{ ccy/hr}$$

$$\frac{489 \text{ ccy/hr}}{0.83} = 589.2 \text{ bcy/hr}$$

Thus,

$$540/589.2 = 0.92 < 1$$

Therefore, only one roller is required to keep up with delivery of material.

Production = 2.5 lcy/cycle × (45 min/hr)/ 0.85 min/cycle = 132 lcy/hr

11.7. d. Bucket fill factor = 85% (average for hard clay from Table 11.6 of *Civil Engineering PE License Review*)

Cycle time = 20 sec (from Table 12.5)

$$\text{Prod.} = \frac{3600 \text{ sec} \times 3 \text{ cy} \times 0.85}{20 \text{ sec/cycle}} \times (45 \text{ min}/60 \text{ min/hr}) \times 1/1.35 = 255 \text{ bcy/hr}$$

11.8 b. Amount to depreciate = $100,000 − $12,000 − $8000 = $80,000

$D_2 = 4/15 \times \$80,000 = \$21,333.33$

Problems 11.9–11.10:

Exhibit 11.9a lists Early Start (ES), Early Finish (EF), Late Start (LS), Late Finish (LF), Total Float (TF), and Free Float (FF) for the diagram. Critical activities are indicated with an asterisk (*).

Activity	Duration	ES	EF	LS	LF	TF	FF
A*	5	0	5	0	5	0	0
B	4	0	4	3	7	3	3
C	1	5	6	6	7	1	1
D*	2	5	7	5	7	0	0
E	2	7	9	8	10	1	0
F*	3	7	10	7	10	0	0

Exhibit 11.9a

To be on the critical path, both total float and free float must be zero.

11.9 c.

11.10 c.

Problems 11.11–11.14.

Critical Activity	a	m	b	t_e	σ_{te}	v
A	6	6	6	6.0	0	0
D	2	5	8	5.0	1.0	1.0
E	2	3	7	3.5	0.83	0.69
F	6	8	10	8.0	0.67	0.45
I	2	2	5	2.5	0.50	0.25

Exhibit 11.11

Sample calculations for activity E:

$$t_e = (a + 4m + b)/6 = (2 + 12 + 7)/6 = 3.5$$
$$\sigma_{te} = (b - a)/6 = (7 - 3)/6 = 0.67$$
$$v = \sigma_{te}^2 = 0.67^2 = 0.45$$

11.11 b. $T_E = 6.0 + 5.0 + 3.5 + 8.0 + 2.5 = 25$ days

11.12 c. $\sigma_{TE} = V^{1/2} = (1.0 + 0.69 + 0.45 + 0.25)^{1/2} = 1.55$

$Z_{-26} = (26 - 25)/1.55 = 0.65$ $P_{-26} = 75\%$ (from Z table)

11.13 a. Before start of day 25 is same as before end of day 24.

$Z_{-24} = (24 - 25)/1.55 = -0.65$ $P_{-24} = 26\%$ (from Z table)

11.14 c. For probability of 93%, $Z = 1.5$.

$1.5 = (X - 25)/1.55 \rightarrow X = 27.35$. Therefore, the project will be completed by the end of day 28 with 93 percent confidence.

REFERENCES

1. Atkins, Harold N. *Highway Materials, Soils, and Concretes.* Prentice Hall, Upper Saddle River, NJ, 1997.
2. Callahan, M. T., Quackenbush, D. G., and Rowings, J. E. *Construction Project Scheduling.* McGraw-Hill, New York, 1992.
3. *Caterpillar Performance Handbook.* Caterpillar. Annual publication.
4. Clough, R. H., and Sears, G. A. *Construction Contracts.* John Wiley and Sons, New York, 1991.
5. Clough, Richard, and Sears, G. A. *Construction Project Management.* John Wiley and Sons, New York, 1991.
6. Evett, J. B. *Surveying.* Prentice Hall, Englewood Cliffs, NJ, 1991.
7. Halpin, D. W. *Construction Management.* John Wiley and Sons, New York, 2006.
8. Halpin, D. W. and Woodhead, R. W. *Construction Management.* John Wiley and Sons, New York, 1998.
9. www.osha.gov/SLTC/trenchingexcavation/index.html
10. Nunnally, S. W. *Construction Methods and Management.* Prentice Hall, Upper Saddle River, NJ, 1998.
11. Parker, H. S., MacGuire, J. W., and Ambrose, J. E. *Simplified Site Engineering.* John Wiley and Sons, New York, 1991.
12. Peurifoy, R. L., Schexnayder, C. J., and Shapira, A. *Construction Planning, Equipment, and Methods.* McGraw-Hill, New York, 2006.
13. Pratt, D. J. *Fundamentals of Construction Estimating.* Delmar Publishers, Albany, NY, 1995.
14. Schroeder, W. L., Dickenson, S. E., and Warrington, D. C. *Soils in Construction.* Prentice Hall, Upper Saddle River, NJ, 2004.

APPENDIX A

Engineering Economics

PROBLEMS

A.1 A loan was made 2½ years ago at 8 percent simple annual interest. The principal amount of the loan has just been repaid along with $600 of interest. The principal amount of the loan was closest to
a. $300
b. $3000
c. $4000
d. $5000

A.2 A $1000 loan was made at 10 percent simple annual interest. It will take how many years for the amount of the loan and interest to equal $1700?
a. 6 years
b. 7 years
c. 8 years
d. 9 years

A.3 A retirement fund earns 8 percent interest, compounded quarterly. If $400 is deposited every three months for 25 years, the amount in the fund at the end of 25 years is nearest to
a. $50,000
b. $75,000
c. $100,000
d. $125,000

A.4 For some interest rate i and some number of interest periods n, the uniform series capital recovery factor is 0.2091, and the sinking fund factor is 0.1941. The interest rate i must be closest to
a. 1½%
b. 2%
c. 3%
d. 4%

A.5 The repair costs for some handheld equipment are estimated to be $120 the first year, increasing by $30 per year in subsequent years. The amount a person will need to deposit into a bank account paying 4 percent interest to provide for the repair costs for the next five years is nearest to
a. $500
b. $600
c. $700
d. $800

A.6 An *annuity* is defined as the
  a. earned interest due at the end of each interest period
  b. cost of producing a product or rendering a service
  c. total annual overhead assigned to a unit of production
  d. series of equal payments occurring at equal periods of time

A.7 One thousand dollars is borrowed for one year at an interest rate of 1 percent per month. If this same sum of money is borrowed for the same period at an interest rate of 12 percent per year, the saving in interest charges is closest to
  a. $0   c. $7
  b. $5   d. $14

A.8 How much should a person invest in a fund that will pay 9 percent, compounded continuously, to have $10,000 in the fund at the end of ten years? The amount is nearest to
  a. $4000   c. $6000
  b. $5000   d. $7000

A.9 A store charges 1½ percent interest per month on credit purchases. This is equivalent to a nominal annual interest rate of
  a. 1.5%   c. 18.0%
  b. 15.0%   d. 19.6%

A.10 A small company borrowed $10,000 to expand its business. The entire principal of $10,000 will be repaid in two years, but quarterly interest of $330 must be paid every three months. The nominal annual interest rate the company is paying is closest to
  a. 3.3%   c. 6.6%
  b. 5.0%   d. 13.2%

A.11 A store policy is to charge 3 percent interest every two months on the unpaid balance in charge accounts. The effective interest rate is closest to
  a. 6%   c. 15%
  b. 12%   d. 19%

A.12 The effective interest rate is 19.56 percent. If there are 12 compounding periods per year, the nominal interest rate is closest to
  a. 1.5%   c. 9.0%
  b. 4.5%   d. 18.0%

A.13 A deposit of $300 was made one year ago into an account paying monthly interest. If the account now has $320.52, the effective annual interest rate is closest to
  a. 7%   c. 12%
  b. 10%   d. 15%

A.14 In a situation where the effective interest rate per year is 12 percent, based on monthly compounding, the nominal interest rate per year is closest to
  a. 8.5%   c. 10.0%
  b. 9.3%   d. 11.4%

A.15 If 10 percent nominal annual interest is compounded daily, the effective annual interest rate is nearest to
  a. 10.00%
  b. 10.38%
  c. 10.50%
  d. 10.75%

A.16 If 10 percent nominal annual interest is compounded continuously, the effective annual interest rate is nearest to
  a. 10.00%
  b. 10.38%
  c. 10.50%
  d. 10.75%

A.17 If the quarterly effective interest rate is 5½ percent with continuous compounding, the nominal interest rate is nearest to
  a. 5.5%
  b. 11.0%
  c. 16.5%
  d. 21.4%

A.18 A continuously compounded loan has what effective interest rate if the nominal interest rate is 25 percent?
  a. $e^{1.25}$
  b. $e^{0.25}$
  c. $e^{0.25} - 1$
  d. $\ln(1.25)$

A.19 A continuously compounded loan has what nominal interest rate if the effective interest rate is 25 percent?
  a. $e^{1.25}$
  b. $e^{0.25}$
  c. $\ln(1.25)$
  d. $\log_{10}(1.25)$

A.20 An individual wishes to deposit a certain quantity of money now to have $500 at the end of five years. With interest at 4 percent per year, compounded semiannually, the amount of the deposit is nearest to
  a. $340
  b. $400
  c. $410
  d. $608

A.21 A steam boiler is purchased on the basis of guaranteed performance. A test indicates that the operating cost will be $300 more per year than the manufacturer guaranteed. If the expected life of the boiler is 20 years and money is worth 8 percent, the amount the purchaser should deduct from the purchase price to compensate for the extra operating cost is nearest to
  a. $2950
  b. $3320
  c. $4100
  d. $5520

A.22 A consulting engineer bought a fax machine with one year's free maintenance. In the second year, the maintenance is estimated at $20. In subsequent years, the maintenance cost will increase $20 per year (that is, third-year maintenance will be $40, fourth-year maintenance will be $60, and so forth). The amount that must be set aside now at 6 percent interest to pay the maintenance costs on the fax machine for the first six years of ownership is nearest to
  a. $101
  b. $164
  c. $229
  d. $284

A.23 An investor is considering buying a 20-year corporate bond. The bond has a face value of $1000 and pays 6 percent interest per year in two semiannual payments. Thus, the purchaser of the bond will receive $30 every six months and, in addition, will receive $1000 at the end of 20 years along with the last $30 interest payment. If the investor seeks to receive 8 percent annual interest, compounded semiannually, the amount the investor is willing to pay for the bond value is closest to
    a. $500   c. $800
    b. $600   d. $900

A.24 Annual maintenance costs for a particular section of highway pavement are $2000. The placement of a new surface would reduce the annual maintenance cost to $500 per year for the first five years and to $1000 per year for the next five years. The annual maintenance after ten years would again be $2000. If maintenance costs are the only savings, the maximum investment that can be justified for the new surface, with interest at 4 percent, is closest to
    a. $5,500   c. $10,000
    b. $7,170   d. $10,340

A.25 A project has an initial cost of $10,000, uniform annual benefits of $2400, and a salvage value of $3000 at the end of its ten-year useful life. At 12 percent interest the net present worth of the project is closest to
    a. $2,500   c. $4,500
    b. $3,500   d. $5,500

A.26 A person borrows $5000 at an interest rate of 18 percent, compounded monthly. Monthly payments of $167.10 are agreed upon. The length of the loan is closest to
    a. 12 months   c. 24 months
    b. 20 months   d. 40 months

A.27 A machine costing $2000 to buy and $300 per year to operate will save labor expenses of $650 per year for eight years. The machine will be purchased if its salvage value at the end of eight years is sufficiently large to make the investment economically attractive. If an interest rate of 10 percent is used, the minimum salvage value must be closest to
    a. $100   c. $300
    b. $200   d. $400

A.28 The amount of money deposited 50 years ago at 8 percent interest that would now provide a perpetual payment of $10,000 per year is nearest to
    a. $3,000   c. $50,000
    b. $8,000   d. $70,000

A.29 An industrial firm must pay a local jurisdiction the cost to expand its sewage treatment plant. In addition, the firm must pay $12,000 annually toward the plant operating costs. The industrial firm will pay sufficient money into a fund that earns 5 percent per year to pay its share of the plant operating costs forever. The amount to be paid to the fund is nearest to
    a. $15,000   c. $160,000
    b. $60,000   d. $240,000

A.30 At an interest rate of 2 percent per month, money will double in value in how many months?
    a. 20 months   c. 50 months
    b. 35 months   d. 65 months

A.31 An individual deposited $10,000 into an account at a credit union. The money was left on deposit for 80 months. During the first 50 months, the deposit earned 12 percent interest, compounded monthly. The credit union then changed its interest policy so that the deposit earned 8 percent interest compounded quarterly during the next 30 months. The amount of money in the account at the end of 80 months is nearest to
    a. $10,000   c. $20,000
    b. $15,000   d. $25,000

A.32 An engineer deposited $200 quarterly into a savings account for three years at 6 percent interest, compounded quarterly. Then for five years, the engineer made no deposits or withdrawals. The amount in the account after eight years is closest to
    a. $1200   c. $2400
    b. $1800   d. $3600

A.33 A sum of money, Q, will be received six years from now. At 6 percent annual interest, the present worth now of Q is $60. At this same interest rate, the value of Q ten years from now is closest to
    a. $60   c. $90
    b. $77   d. $107

A.34 If $200 is deposited in a savings account at the beginning of each of 15 years and the account earns interest at 6 percent, compounded annually, the value of the account at the end of 15 years will be most nearly
    a. $4500   c. $4900
    b. $4700   d. $5100

A.35 The maintenance expense on a piece of machinery is estimated as follows:

Year	1	2	3	4
Maintenance	$150	$300	$450	$600

If interest is 8 percent, the equivalent uniform annual maintenance cost is closest to
    a. $250   c. $350
    b. $300   d. $400

A.36 A payment of $12,000 six years from now is equivalent, at 10 percent interest, to an annual payment for eight years starting at the end of this year. The annual payment is closest to
    a. $1000   c. $1400
    b. $1200   d. $1600

A.37 A manufacturer purchased $15,000 worth of equipment with a useful life of six years and a $2000 salvage value at the end of the six years. Assuming a 12 percent interest rate, the equivalent uniform annual cost is nearest to
  a. $1500
  b. $2500
  c. $3500
  d. $4500

A.38 Consider a machine as follows:

Initial cost: $80,000

End-of-useful-life salvage value: $20,000

Annual operating cost: $18,000

Useful life: 20 years

Based on 10 percent interest, the equivalent uniform annual cost for the machine is closest to
  a. $21,000
  b. $23,000
  c. $25,000
  d. $27,000

A.39 Consider a machine as follows:

Initial cost: $80,000

Annual operating cost: $18,000

Useful life: 20 years

What must be the salvage value of the machine at the end of 20 years for the machine to have an equivalent uniform annual cost of $27,000? Assume a 10 percent interest rate. The salvage value is closest to
  a. $10,000
  b. $20,000
  c. $40,000
  d. $50,000

A.40 Twenty-five thousand dollars is deposited in a savings account that pays 5 percent interest, compounded semiannually. Equal annual withdrawals are to be made from the account beginning one year from now and continuing forever. The maximum amount of the equal annual withdrawals is closest to
  a. $625
  b. $1000
  c. $1250
  d. $1265

A.41 An investor is considering investing $10,000 in a piece of land. The property taxes are $100 per year. The lowest selling price the investor must receive to earn a 10 percent interest rate after keeping the land for ten years is
  a. $21,000
  b. $23,000
  c. $27,000
  d. $31,000

A.42 The rate of return of a $10,000 investment that will yield $1000 per year for 20 years is closest to
  a. 1%
  b. 4%
  c. 8%
  d. 12%

A.43 An engineer invested $10,000 in a company, in return receiving $600 per year for six years and the $10,000 investment back at the end of the six years. The rate of return on the investment was closest to
    a. 6%
    b. 10%
    c. 12%
    d. 15%

A.44 An engineer made ten annual end-of-year purchases of $1000 of common stock. At the end of the tenth year, just after the last purchase, the engineer sold all the stock for $12,000. The rate of return received on the investment is closest to
    a. 2%
    b. 4%
    c. 8%
    d. 10%

A.45 A company is considering buying a new piece of machinery.

Initial cost: $80,000

End-of-useful-life salvage value: $20,000

Annual operating cost: $18,000

Useful life: 20 years

The machine will produce an annual savings in material of $25,700. What is the before-tax rate of return if the machine is installed? The rate of return is closest to
    a. 6%
    b. 8%
    c. 10%
    d. 15%

A.46 Consider the following situation: invest $100 now and receive two payments of $102.15—one at the end of Year 3 and one at the end of Year 6. The rate of return is nearest to
    a. 8%
    b. 12%
    c. 18%
    d. 22%

A.47 Two mutually exclusive alternatives are being considered:

Year	A	B
0	−$2500	−$6000
1	+746	+1664
2	+746	+1664
3	+746	+1664
4	+746	+1664
5	+746	+1664

The rate of return on the difference between the alternatives is closest to
    a. 6%
    b. 8%
    c. 10%
    d. 12%

A.48 A project will cost $50,000. The benefits at the end of the first year are estimated to be $10,000, increasing $1000 per year in subsequent years. Assuming a 12 percent interest rate, no salvage value, and an eight-year analysis period, the benefit-cost ratio is closest to
a. 0.78
b. 1.00
c. 1.28
d. 1.45

A.49 Two alternatives are being considered:

	A	B
Initial cost	$500	$800
Uniform annual benefit	$140	$200
Useful life, years	8	8

The benefit-cost ratio of the difference between the alternatives, based on a 12 percent interest rate, is closest to
a. 0.60
b. 0.80
c. 1.00
d. 1.20

A.50 An engineer will invest in a mining project if the benefit-cost ratio is greater than 1.0, based on an 18 percent interest rate. The project cost is $57,000. The net annual return is estimated at $14,000 for each of the next eight years. At the end of eight years, the mining project will be worthless. The benefit-cost ratio is closest to
a. 1.00
b. 1.05
c. 1.21
d. 1.57

A.51 A city has retained your firm to do a benefit-cost analysis of the following project:

Project cost: $60,000,000

Gross income: $20,000,000 per year

Operating costs: $5,500,000 per year

Salvage value after ten years: None

The project life is ten years. Use 8 percent interest in the analysis. The computed benefit-cost ratio is closest to
a. 0.80
b. 1.00
c. 1.50
d. 1.60

A.52 A piece of property is purchased for $10,000 and yields a $1000 yearly profit. If the property is sold after five years, the minimum price to break even, with interest at 6 percent, is closest to
a. $5000
b. $6500
c. $7700
d. $8300

A.53 Given two machines:

	A	B
Initial cost	$55,000	$75,000
Total annual costs	$16,200	$12,450

With interest at 10 percent per year, at what service life do these two machines have the same equivalent uniform annual cost? The service life is closest to
- a. 5 years
- b. 6 years
- c. 7 years
- d. 8 years

A.54 A machine part that is operating in a corrosive atmosphere is made of low-carbon steel. It costs $350 installed and lasts six years. If the part is treated for corrosion resistance, it will cost $700 installed. How long must the treated part last to be as economic as the untreated part, if money is worth 6 percent?
- a. 8 years
- b. 11 years
- c. 15 years
- d. 17 years

A.55 A firm has determined the two best paints for its machinery are Tuff-Coat at $45 per gallon and Quick at $22 per gallon. The Quick paint is expected to prevent rust for five years. Both paints take $40 of labor per gallon to apply, and both cover the same area. If a 12 percent interest rate is used, how long must the Tuff-Coat paint prevent rust to justify its use?
- a. 5 years
- b. 6 years
- c. 7 years
- d. 8 years

A.56 Two alternatives are being considered:

	A	B
Cost	$1000	$2000
Useful life in years	10	10
End-of-useful-life salvage value	100	400

The net annual benefit of A is $150. If interest is 8 percent, what must be the net annual benefit of B for the two alternatives to be equally desirable? The net annual benefit of B must be closest to
- a. $150
- b. $200
- c. $225
- d. $275

A.57 Which one of the following is NOT a method of depreciating plant equipment for accounting and engineering economics purposes?
- a. double-entry method
- b. modified accelerated cost recovery system
- c. sum-of-years-digits method
- d. straight-line method

A.58 A machine costs $80,800, has a 20-year useful life, and an estimated $20,000 end-of-useful-life salvage value. Assuming sum-of-years-digits depreciation, the book value of the machine after two years is closest to
   a. $21,000
   b. $42,000
   c. $59,000
   d. $69,000

A.59 A machine costs $100,000. After its 25-year useful life, its estimated salvage value is $5,000. Based on double-declining-balance depreciation, what will be the book value of the machine at the end of three years? The book value is closest to
   a. $16,000
   b. $22,000
   c. $58,000
   d. $78,000

Questions A.60 to A.63

Special tools for the manufacture of finished plastic products cost $15,000 and have an estimated $1000 salvage value at the end of an estimated three-year useful life.

A.60 The third-year straight-line depreciation is closest to
   a. $3000
   b. $3500
   c. $4000
   d. $4500

A.61 The first-year modified-accelerated-cost-recovery-system (MACRS) depreciation is closest to
   a. $3000
   b. $4000
   c. $5000
   d. $6000

A.62 The second-year sum-of-years-digits (SOYD) depreciation is closest to
   a. $3000
   b. $3500
   c. $4000
   d. $4500

A.63 The second-year sinking-fund depreciation, based on 8 percent interest, is nearest to
   a. $3000
   b. $3500
   c. $4000
   d. $4500

A.64 An individual who has a 28 percent incremental income tax rate is considering purchasing a $1000 taxable corporation bond. The bondholder will receive $100 a year in interest and the original $1000 back when the bond becomes due in six years. This individual's after-tax rate of return from the bond is nearest to
   a. 6%
   b. 7%
   c. 8%
   d. 9%

A.65 A $20,000 investment in equipment will produce $6000 of net annual benefits for the next eight years. The equipment will be depreciated by straight-line depreciation over its eight-year useful life. The equipment has no salvage value. Assuming a 34 percent income tax rate, the after-tax rate of return for this investment is closest to
   a. 8%
   b. 10%
   c. 12%
   d. 18%

A.66 An individual bought a one-year savings certificate for $10,000, and it pays 6 percent. This person's taxable income renders a 28 percent incremental income tax rate. The after-tax rate of return on this investment is closest to
   a. 2%
   b. 3%
   c. 4%
   d. 5%

A.67 A tool costing $300 has no salvage value. Its resulting before-tax cash flow is shown in the following partially completed cash flow table.

Year	Before-Tax Cash Flow	Effect on SOYD Deprec	Effect on Taxable Income	Income Taxes	After-Tax Cash Flow
0	−$300				
1	+100				
2	+150				
3	+200				

The tool is to be depreciated over three years using sum-of-years-digits depreciation. The income tax rate is 50 percent. The after-tax rate of return is nearest to
   a. 8%
   b. 10%
   c. 12%
   d. 15%

A.68 An engineer is considering the purchase of an annuity that will pay $1000 per year for ten years. The engineer desires a 5 percent rate of return on the annuity after considering the effect of an estimated 6 percent inflation per year. The amount the engineer would be willing to pay for the annuity is closest to
   a. $1500
   b. $3000
   c. $4500
   d. $6000

A.69 An automobile costs $20,000 today. You can earn 12 percent tax-free on an "auto purchase account." If you expect the cost of the auto to increase by 10 percent per year, the amount you would need to deposit in the account to provide for the purchase of the auto five years from now is closest to
   a. $12,000
   b. $14,000
   c. $16,000
   d. $18,000

A.70 An engineer purchases a building lot for $40,000 cash and plans to sell it after five years. To deliver an 18 percent before-tax rate of return, after taking the 6 percent annual inflation rate into account, the selling price must be nearest to
   a. $100,000
   b. $125,000
   c. $150,000
   d. $175,000

A.71 A piece of equipment with a list price of $450 can actually be purchased for either $400 cash or $50 immediately plus four additional annual payments of $115.25. All values are in dollars of current purchasing power. If the typical customer considered a 5 percent interest rate appropriate, the inflation rate at which the two purchase alternatives are equivalent is nearest to

a. 5%   c. 8%
b. 6%   d. 10%

A.72 A man wants to determine whether to invest $1000 in a friend's speculative venture. He will do so if he thinks he can get this money back. The probabilities of the various outcomes at the end of one year are:

Result	Probability
$2000 (double his money)	0.3
1500	0.1
1000	0.2
500	0.3
0 (lose everything)	0.1

His expected outcome if he invests the $1000 is closest to

a. $800   c. $1000
b. $900   d. $1100

A.73 The amount you would be willing to pay for an insurance policy protecting you against a 1 in 20 chance of losing $10,000 three years from now, if interest is 10 percent, is closest to

a. $175   c. $1000
b. $350   d. $1500

SOLUTIONS

A.1 **b.**

$$F = P + Pin$$
$$600 + P = P + P(0.08)(2.50)$$
$$P = [600]/[0.08(2.50)] = \$3000$$

A.2 **b.**

$$F = P + Pin$$
$$1700 = 1000 + 1000(0.10)(n)$$
$$n = [700]/[1000(0.10)] = 7 \text{ years}$$

A.3 **d.**

$$F = A(F/A, i, n) = 400(F/A, 2\%, 100)$$
$$= 400(312.33) = \$124{,}890$$

A.4 **a.** The relationship between the capital recovery factor and the sinking fund factor is $(A/P,i,n) = (A/F,i,n) + i$. Substituting the values in the problem

$$0.2091 = 0.1941 + i$$
$$i = 0.2091 - 0.1941 = 0.015 = 1\frac{1}{2}\%$$

A.5 **d.**

$$\begin{aligned}P &= A(P/A,i,n) + G(P/G,i,n)\\ &= 120(P/A,4\%,5) + 30(P/G,4\%,5)\\ &= 120(4.452) + 30(8.555) = \$791\end{aligned}$$

A.6 **d.**

A.7 **c.**

At $i = 1\%$/month: $F = 1000(1 + 0.01)^{12} = \$1126.83$

At $i = 12\%$/year: $F = 1000(1 + 0.12)^{1} = \$1120.00$

Saving in interest charges = $1126.83 - 1120.00 = \$6.83$

A.8 **a.**

$$P = Fe^{-rn} = 10,000e^{-0.09(10)} = 4066$$

A.9 **c.** The nominal interest rate is the annual interest rate ignoring the effect of any compounding. Nominal interest rate = $1\frac{1}{2}\% \times 12 = 18\%$.

A.10 **d.** The interest paid per year = $330 \times 4 = 1320$. The nominal annual interest rate = $1320/10,000 = 0.132 \approx 13.2\%$.

A.11 **d.**

$$i_{\text{eff}} = (1 + r/m)^m - 1 = (1 + 0.03)^6 - 1 = 0.194 = 19.4\%$$

A.12 **d.**

$$i_{\text{eff}} = (1 + r/m)^m - 1$$
$$r/m = (1 + i_{\text{eff}})^{1/m} - 1 = (1 + 0.1956)^{1/12} = 0.015$$
$$r = 0.015(m) = 0.015 \times 12 = 0.18 = 18\%$$

A.13 **a.**

$$i_{\text{eff}} = 20.52/300 = 0.0684 = 6.84\%$$

A.14 **d.**

$$i_{\text{eff}} = (1 + r/m)^m - 1$$
$$0.12 = (1 + r/12)^{12} - 1$$
$$(1.12)^{1/12} = (1 + r/12)$$
$$1.00949 = (1 + r/12)$$
$$r = 0.00949 \times 12 = 0.1138 = 11.38\%$$

A.15 c.
$$i_{eff} = (1 + r/m)^m - 1 = (1 + 0.10/365)^{365} - 1 = 0.1052 = 10.52\%$$

A.16 c.
$$i_{eff} = e^r - 1$$
where r = nominal annual interest rate
$$i_{eff} = e^{0.10} - 1 = 0.10517 = 10.52\%$$

A.17 d. For 3 months: $i_{eff} = e^r - 1$; $0.055 = e^r - 1$
The rate per quarter year is $r = \ln(1.055) = 0.05354$; $r = 4 \times 0.05354$
$= 0.214 = 21.4\%$ per year

A.18 c.
$$i_{eff} = e^r - 1 = e^{0.25} - 1$$

A.19 c.
$$i_{eff} = e^r - 1 = 0.25$$
$$e^r = 1.25$$
$$\ln(e^r) = \ln(1.25)$$
$$r = \ln(1.25)$$

A.20 c.
$$P = F(P/F, i, n) = 500(P/F, 2\%, 10) = 500(0.8203) = \$410$$

A.21 a.
$$P = 300(P/A, 8\%, 20) = 300(9.818) = \$2945$$

A.22 c. Using single payment present worth factors:
$$P = 20(P/F, 6\%, 2) + 40(P/F, 6\%, 3) + 60(P/F, 6\%, 4)$$
$$+ 80(P/F, 6\%, 5) + 100(P/F, 6\%, 6) = \$229$$

Alternate solution using the gradient present worth factor:
$$P = 20(P/G, 6\%, 6) = 20(11.459) = \$229$$

A.23 c.
$$PW = 30(P/A, 4\%, 40) + 1000(P/F, 4\%, 40)$$
$$= 30(19.793) + 1000(0.2083) = \$802$$

A.24 d. Benefits are $1500 per year for the first five years and $1000 per year for the subsequent five years.
As Exhibit A.24 indicates, the benefits may be considered as $1000 per year for ten years plus an additional $500 benefit in each of the first five years.

$$\text{Maximum investment} = \text{Present worth of benefits}$$
$$= 1000(P/A, 4\%, 10) + 500(P/A, 4\%, 5)$$
$$= 1000(8.111) + 500(4.452) = \$10,337$$

Exhibit A.24

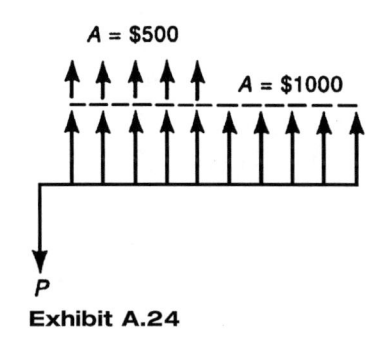

A.25 c.

$$\text{NPW} = \text{PW of benefits} - \text{PW of cost}$$
$$= 2400(P/A, 12\%, 10) + 3000(P/F, 12\%, 10) - 10,000 = \$4526$$

A.26 d.

$$\text{PW of benefits} = \text{PW of cost}$$
$$5000 = 167.10(P/A, 1.5\%, n)$$
$$(P/A, 1.5\%, n) = 5000/167.10 = 29.92$$

From the $1\frac{1}{2}\%$ interest table, $n = 40$.

A.27 c.

$$\text{NPW} = \text{PW of benefits} - \text{PW of cost} = 0$$
$$= (650 - 300)(P/A, 10\%, 8) + S_8(P/F, 10\%, 8) - 2000 = 0$$
$$S_8 = 132.75/0.4665 = \$285$$

A.28 a. The amount of money needed now to begin the perpetual payments is $P' = A/i = 10,000/0.08 = 125,000$. From this, we can compute the amount of money, P, that would need to have been deposited 50 years ago:

$$P = 125,000(P/F, 8\%, 50) = 125,000(0.0213) = \$2663$$

A.29 d.

$$P = A/i = 12,000/0.05 = \$240,000$$

A.30 b.

$$2 = 1(F/P, i, n)$$
$$(F/P, 2\%, n) = 2$$

From the 2% interest table, $n =$ about 35 months.

A.31 c. At end of 50 months,

$$F = 10,000(F/P, 1\%, 50) = 10,000(1.645) = \$16,450$$

At end of 80 months,

$$F = 16,450(F/P, 2\%, 10) = 16,450(1.219) = \$20,053$$

A.32 **d.**
$$FW = 200(F/A, 1.5\%, 12)(F/P, 1.5\%, 20)$$
$$= 200(13.041)(1.347) = \$3513$$

A.33 **d.** The present sum $P = 60$ is equivalent to Q six years hence at 6 percent interest. The future sum F may be calculated by either of two methods:
$$F = Q(F/P, 6\%, 4) \quad \text{and} \quad Q = 60(F/P, 6\%, 6)$$
$$F = P(F/P, 6\%, 10)$$

Since P is known, the second equation may be solved directly.
$$F = P(F/P, 6\%, 10) = 60(1.791) = \$107$$

A.34 **c.**
$$F' = A(F/A, i, n) = 200(F/A, 6\%, 15) = 200(23.276) = \$4655.20$$
$$F = F'(F/P, i, n) = 4655.20(F/P, 6\%, 1) = 4655.20(1.06) = \$4935$$

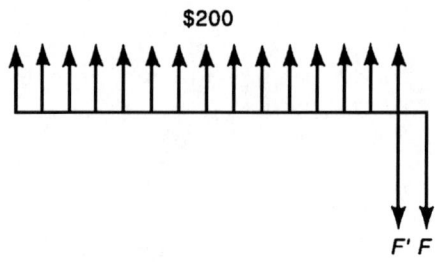

Exhibit A.34

A.35 **c.**
$$EUAC = 150 + 150(A/G, 8\%, 4) = 150 + 150(1.404) = \$361$$

A.36 **b.**
$$\text{Annual payment} = 12{,}000(P/F, 10\%, 6)(A/P, 10\%, 8)$$
$$= 12{,}000(0.5645)(0.1874) = \$1269$$

A.37 **c.**
$$EUAC = 15{,}000(A/P, 12\%, 6) - 2000(A/F, 12\%, 6)$$
$$= 15{,}000(0.2432) - 2000(0.1232) = \$3402$$

A.38 **d.**
$$EUAC = 80{,}000(A/P, 10\%, 20) - 20{,}000(A/F, 10\%, 20)$$
$$+ \text{annual operating cost}$$
$$= 80{,}000\,(0.1175) - 20{,}000\,(0.0175) + 18{,}000$$
$$= 9400 - 350 + 18{,}000 = \$27{,}050$$

A.39 **b.**
$$EUAC = EUAB$$
$$27{,}000 = 80{,}000(A/P, 10\%, 20) + 18{,}000 - S(A/F, 10\%, 20)$$
$$= 80{,}000(0.1175) + 18{,}000 - S(0.0175)$$
$$S = (27{,}400 - 27{,}000)/0.0175 = \$22{,}857$$

A.40 d. The general equation for an infinite life, $P = A/i$, must be used to solve the problem.

$$i_{\text{eff}} = (1 + 0.025)^2 - 1 = 0.050625$$

The maximum annual withdrawal will be $A = Pi = 25{,}000(0.050625) = \$1266$.

A.41 c.

$$\begin{aligned}\text{Minimum sale price} &= 10{,}000(F/P, 10\%, 10) + 100(F/A, 10\%, 10) \\ &= 10{,}000(2.594) + 100(15.937) = \$27{,}530\end{aligned}$$

A.42 c.

$$\text{NPW} = 1000(P/A, i, 20) - 10{,}000 = 0$$
$$(P/A, i, 20) = 10{,}000/1000 = 10$$

From interest tables: $6\% < i < 8\%$.

A.43 a. The rate of return was $= 600/10{,}000 = 0.06 = 6\%$.

A.44 b.

$$F = A(F/A, i, n)$$
$$12{,}000 = 1000(F/A, i, 10)$$
$$(F/A, i, 10) = 12{,}000/1000 = 12$$

In the 4% interest table: $(F/A, 4\%, 10) = 12.006$, so $i = 4\%$.

A.45 b.

PW of cost = PW of benefits
$$80{,}000 = (25{,}700 - 18{,}000)(P/A, i, 20) + (20{,}000(P/F, i, 20)$$

Try $i = 8\%$.

$$80{,}000 = 7709(9.818) + 20{,}000(0.2145) = 79{,}889$$

Therefore, the rate of return is very close to 8 percent.

A.46 c.

PW of cost = PW of benefits
$$100 = 102.15(P/F, i, 3) + 102.15(P/F, i, 6)$$

Solve by trial and error:
Try $i = 12\%$.
$$100 = 102.15(0.7118) + 102.15(0.5066) = 124.46$$

The PW of benefits exceeds the PW of cost. This indicates that the interest rate i is too low. Try $i = 18\%$.
$$100 = 102.15(0.6086) + 102.15(0.3704) = 100.00$$

Therefore, the rate of return is 18 percent.

A.47 c. The difference between the alternatives:

Incremental cost = 6000 − 2500 = $3500

Incremental annual benefit = 1664 − 746 = $918

$$PW \text{ of cost} = PW \text{ of benefits}$$
$$3500 = 918(P/A, i, 5)$$
$$(P/A, i, 5) = 3500/918 = 3.81$$

From the interest tables, i is very close to 10 percent.

A.48 c.

$$B/C = \frac{PW \text{ of benefits}}{PW \text{ of cost}} = \frac{10,000(P/A, 12\%, 8) + 1000(P/G, 12\%, 8)}{50,000}$$
$$= \frac{10,000(4.968) + 1000(14.471)}{50,000} = 1.28$$

A.49 c.

$$B/C = \frac{PW \text{ of benefits}}{PW \text{ of cost}} = \frac{60(P/A, 12\%, 8)}{300} = \frac{60(4.968)}{300} = 0.99$$

Alternate solution:

$$B/C = \frac{EUAB}{EUAC} = \frac{60}{300(A/P, 12\%, 8)} = \frac{60}{300(0.2013)} = 0.99$$

A.50 a.

$$B/C = \frac{PW \text{ of benefits}}{PW \text{ of cost}} = \frac{14,000(P/A, 18\%, 8)}{57,000} = \frac{14,000(4.078)}{57,000} = 1.00$$

A.51 d.

$$B/C = \frac{EUAB}{EUAC} = \frac{20,000,000 - 5,500,000}{60,000,000(A/P, 8\%, 10)} = 1.62$$

A.52 c.

$$F = 10,000(F/P, 6\%, 5) - 1000(F/A, 6\%, 5)$$
$$= 10,000(1.338) - 1000(5.637) = \$7743$$

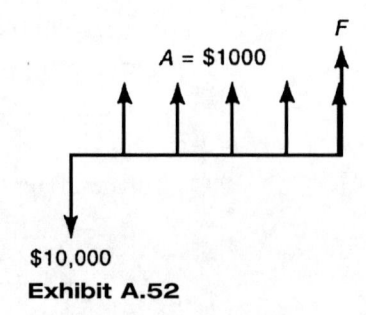

Exhibit A.52

A.53 d.

$$\text{PW of cost}_A = \text{PW of cost}_B$$
$$55{,}000 + 16{,}200(P/A, 10\%, n) = 75{,}000 + 12{,}450(P/A, 10\%, n)$$
$$(P/A, 10\%, n) = (75{,}000 - 55{,}000)/(16{,}200 - 12{,}450)$$
$$= 5.33$$

From the 10% interest table, $n = 8$ years.

A.54 c.

$$\text{EUAC}_{\text{untreated}} = \text{EUAC}_{\text{treated}}$$
$$350(A/P, 6\%, 6) = 700(A/P, 6\%, n)$$
$$350(0.2034) = 700(A/P, 6\%, n)$$
$$(A/P, 6\%, n) = 71.19/700 = 0.1017$$

From the 6% interest table, $n = 15+$ years.

A.55 d.

$$\text{EUAC}_{T-C} = \text{EUAC}_{\text{Quick}}$$
$$(45 + 40)(A/P, 12\%, n) = (22 + 40)(A/P, 12\%, 5)$$
$$(A/P, 12\%, n) = 17.20/85 = 0.202$$

From the 12% interest table, $n = 8$.

A.56 d. At breakeven,

$$\text{NPW}_A = \text{NPW}_B$$
$$150(P/A, 8\%, 10) + 100(P/F, 8\%, 10) - 1000 = \text{NAB}(P/A, 8\%, 10)$$
$$+ 400(P/F, 8\%, 10) - 2000$$
$$52.82 = 6.71(\text{NAB}) - 1814.72$$

Net annual benefit (NAB) = $(1814.72 + 52.82)/6.71 = \$278$.

A.57 a. *Double entry* probably is a reference to double entry accounting. It is not a method of depreciation.

A.58 d. Sum-of-years-digits depreciation:

$$D_j = \frac{n - j + 1}{\frac{n}{2}(n+1)}(C - S_n)$$

$$D_1 = \frac{20 - 1 + 1}{\frac{20}{2}(20+1)}(80{,}000 - 20{,}000) = 5714$$

$$D_2 = \frac{20 - 2 + 1}{\frac{20}{2}(20+1)}(80{,}000 - 20{,}000) = 5429$$

Total: $11,143

Book value = Cost − Depreciation to date
= 80,000 − 11,143 = $68,857

A.59 d. Double-declining-balance depreciation:

$$BV_j = C\left(1 - \frac{2}{n}\right)^j$$

$$BV_3 = 100{,}000\left(1 - \frac{2}{25}\right)^3 = \$77{,}869$$

A.60 d. $D_3 = (C - S)/n = (15{,}000 - 1000)/3 = \$4666$

A.61 c. The half-year convention applies here; double-declining balance must be used with an assumed salvage value of zero. In general,

$$D_j = \frac{2C}{n}\left(1 - \frac{2}{n}\right)^{j-1}$$

For a half-year in Year 1:

$$D_1 = \frac{1}{2} \times \frac{2 \times 15{,}000}{3}\left(1 - \frac{2}{3}\right)^{1-1} = \$5000$$

A.62 d.

$$D_j = \frac{n - j + 1}{\frac{n}{2}(n+1)}(C - S_n)$$

$$D_2 = \frac{3 - 2 + 1}{\frac{3}{2}(3+1)}(15{,}000 - 1000) = \$4667$$

A.63 d.

$$D_2 = (15{,}000 - 1000)(A/F, 8\%, 3)(F/P, 8\%, 1)$$
$$= 14{,}000(0.3080)(1.08) = \$4657$$

A.64 b. Twenty-eight percent of the $100 interest income must be paid in taxes. The balance of $72 is the after-tax income. Thus, the after-tax rate of return = 72/1000 = 0.072 = 7.2%.

A.65 d.

Year	Before-Tax Cash Flow	SL Deprec	Taxable Income	34% Income Taxes	After-Tax Cash Flow
0	−$20,000				−$20,000
1–8	+6000	2500	3500	1190	+4810

$$D_j = (P - S)/n = \frac{20{,}000 - 0}{8} = 2500$$

PW of cost = PW of benefits
$20{,}000 = 4810(P/A, i, 8)$
$(P/A, i, 8) = 20{,}000/4810 = 4.158$

From interest tables, $i = 18\%$.

A.66 c. Additional taxable income = 0.06(10,000) = 600

Additional income tax = 0.28(600) = 168

After-tax rate income = 600 − 168 = 432

After-tax rate of return = 432/10,000 = 0.043 = 4.3%

Alternate solution:

After-tax rate of return = (1 − Incremental tax rate)(Before-tax rate of return)
= (1 − 0.28)(0.06) = 0.043 = 4.3%

A.67 c.

Year	Before-Tax Cash Flow	Effect on SOYD Deprec	Effect on Taxable Income	50% Income Taxes	After-Tax Cash Flow
0	−$300				−$300
1	+100	−150	−50	+25	+125
2	+150	−100	+50	−25	+125
3	+200	−50	+150	−75	+125

For the after-tax cash flow:

$$PW \text{ of cost} = PW \text{ of benefits}$$
$$300 = 125(P/A, i, 3)$$
$$(P/A, i, 3) = 300/125 = 2.40$$

From the interest table, we find that i (the after-tax rate of return) is close to 12.

A.68 d.

$$d = i + f + if = 0.05 + 0.06 + 0.05(0.06) = 0.113 = 11.3\%$$
$$P = A(P/A, 11.3\%, 10) = 100 \left[\frac{(1+0.113)^{10} - 1}{0.113(1+0.113)^{10}} \right]$$
$$= 1000 \left[\frac{1.9171}{0.3296} \right] = \$5816$$

A.69 d.

Cost of auto five years hence $(F) = P(1 + \text{inflation rate})^n$
$$= 20,000(1 + 0.10)^5 = 32,210$$

Amount to deposit now to have $32,210 available five years hence:

$$P = F(P/F, i, n) = 32,210(P/F, 12\%, 5) = 32,210(0.5674) = \$18,276$$

A.70 b.

Selling price $(F) = 40,000(F/P, 18\%, 5)(F/P, 6\%, 5)$
$$= 40,000(2.288)(1.338) = \$122,500$$

A.71 b.

$$\text{PW of cash purchase} = \text{PW of installment purchase}$$
$$400 = 50 + 115.25(P/A,d,4)$$
$$(P/A,d,4) = 350/115.25 = 3.037$$

From the interest tables, $d = 12\%$.

$$d = i + f + i(f)$$
$$0.12 = 0.05 + f + 0.05f$$
$$f = 0.07/1.05 = 0.0667 = 6.67\%$$

A.72 d. The expected income $= 0.3(2000) + 0.1(1500) + 0.2(1000) + 0.3(500) + 0.1(0) = \$1100$.

A.73 b.

$$\text{PW of benefit} = (10{,}000/20)(P/F,10\%,3)$$
$$= 500(0.7513) = \$376$$

Compound interest factors

$\frac{1}{2}$%

	Single Payment		Uniform Payment Series				Uniform Gradient		
	Compound Amount Factor	Present Worth Factor	Sinking Fund Factor	Capital Recovery Factor	Compound Amount Factor	Present Worth Factor	Gradient Uniform Series	Gradient Present Worth	
n	Find F Given P F/P	Find P Given F P/F	Find A Given F A/F	Find A Given P A/P	Find F Given A F/A	Find P Given A P/A	Find A Given G A/G	Find P Given G P/G	n
1	1.005	.9950	1.0000	1.0050	1.000	0.995	0	0	1
2	1.010	.9901	.4988	.5038	2.005	1.985	0.499	0.991	2
3	1.015	.9851	.3317	.3367	3.015	2.970	0.996	2.959	3
4	1.020	.9802	.2481	.2531	4.030	3.951	1.494	5.903	4
5	1.025	.9754	.1980	.2030	5.050	4.926	1.990	9.803	5
6	1.030	.9705	.1646	.1696	6.076	5.896	2.486	14.660	6
7	1.036	.9657	.1407	.1457	7.106	6.862	2.980	20.448	7
8	1.041	.9609	.1228	.1278	8.141	7.823	3.474	27.178	8
9	1.046	.9561	.1089	.1139	9.182	8.779	3.967	34.825	9
10	1.051	.9513	.0978	.1028	10.228	9.730	4.459	43.389	10
11	1.056	.9466	.0887	.0937	11.279	10.677	4.950	52.855	11
12	1.062	.9419	.0811	.0861	12.336	11.619	5.441	63.218	12
13	1.067	.9372	.0746	.0796	13.397	12.556	5.931	74.465	13
14	1.072	.9326	.0691	.0741	14.464	13.489	6.419	86.590	14
15	1.078	.9279	.0644	.0694	15.537	14.417	6.907	99.574	15
16	1.083	.9233	.0602	.0652	16.614	15.340	7.394	113.427	16
17	1.088	.9187	.0565	.0615	17.697	16.259	7.880	128.125	17
18	1.094	.9141	.0532	.0582	18.786	17.173	8.366	143.668	18
19	1.099	.9096	.0503	.0553	19.880	18.082	8.850	160.037	19
20	1.105	.9051	.0477	.0527	20.979	18.987	9.334	177.237	20
21	1.110	.9006	.0453	.0503	22.084	19.888	9.817	195.245	21
22	1.116	.8961	.0431	.0481	23.194	20.784	10.300	214.070	22
23	1.122	.8916	.0411	.0461	24.310	21.676	10.781	233.680	23
24	1.127	.8872	.0393	.0443	25.432	22.563	11.261	254.088	24
25	1.133	.8828	.0377	.0427	26.559	23.446	11.741	275.273	25
26	1.138	.8784	.0361	.0411	27.692	24.324	12.220	297.233	26
27	1.144	.8740	.0347	.0397	28.830	25.198	12.698	319.955	27
28	1.150	.8697	.0334	.0384	29.975	26.068	13.175	343.439	28
29	1.156	.8653	.0321	.0371	31.124	26.933	13.651	367.672	29
30	1.161	.8610	.0310	.0360	32.280	27.794	14.127	392.640	30
36	1.197	.8356	.0254	.0304	39.336	32.871	16.962	557.564	36
40	1.221	.8191	.0226	.0276	44.159	36.172	18.836	681.341	40
48	1.270	.7871	.0185	.0235	54.098	42.580	22.544	959.928	48
50	1.283	.7793	.0177	.0227	56.645	44.143	23.463	1035.70	50
52	1.296	.7716	.0169	.0219	59.218	45.690	24.378	1113.82	52
60	1.349	.7414	.0143	.0193	69.770	51.726	28.007	1448.65	60
70	1.418	.7053	.0120	.0170	83.566	58.939	32.468	1913.65	70
72	1.432	.6983	.0116	.0166	86.409	60.340	33.351	2012.35	72
80	1.490	.6710	.0102	.0152	98.068	65.802	36.848	2424.65	80
84	1.520	.6577	.00961	.0146	104.074	68.453	38.576	2640.67	84
90	1.567	.6383	.00883	.0138	113.311	72.331	41.145	2976.08	90
96	1.614	.6195	.00814	.0131	122.829	76.095	43.685	3324.19	96
100	1.647	.6073	.00773	.0127	129.334	78.543	45.361	3562.80	100
104	1.680	.5953	.00735	.0124	135.970	80.942	47.025	3806.29	104
120	1.819	.5496	.00610	.0111	163.880	90.074	53.551	4823.52	120
240	3.310	.3021	.00216	.00716	462.041	139.581	96.113	13,415.56	240
360	6.023	.1660	.00100	.00600	1004.5	166.792	128.324	21,403.32	360
480	10.957	.0913	.00050	.00550	1991.5	181.748	151.795	27,588.37	480

Compound interest factors

1%

	Single Payment		Uniform Payment Series				Uniform Gradient		
	Compound Amount Factor	Present Worth Factor	Sinking Fund Factor	Capital Recovery Factor	Compound Amount Factor	Present Worth Factor	Gradient Uniform Series	Gradient Present Worth	
n	Find F Given P F/P	Find P Given F P/F	Find A Given F A/F	Find A Given P A/P	Find F Given A F/A	Find P Given A P/A	Find A Given G A/G	Find P Given G P/G	n
1	1.010	.9901	1.0000	1.0100	1.000	0.990	0	0	1
2	1.020	.9803	.4975	.5075	2.010	1.970	0.498	0.980	2
3	1.030	.9706	.3300	.3400	3.030	2.941	0.993	2.921	3
4	1.041	.9610	.2463	.2563	4.060	3.902	1.488	5.804	4
5	1.051	.9515	.1960	.2060	5.101	4.853	1.980	9.610	5
6	1.062	.9420	.1625	.1725	6.152	5.795	2.471	14.320	6
7	1.072	.9327	.1386	.1486	7.214	6.728	2.960	19.917	7
8	1.083	.9235	.1207	.1307	8.286	7.652	3.448	26.381	8
9	1.094	.9143	.1067	.1167	9.369	8.566	3.934	33.695	9
10	1.105	.9053	.0956	.1056	10.462	9.471	4.418	41.843	10
11	1.116	.8963	.0865	.0965	11.567	10.368	4.900	50.806	11
12	1.127	.8874	.0788	.0888	12.682	11.255	5.381	60.568	12
13	1.138	.8787	.0724	.0824	13.809	12.134	5.861	71.112	13
14	1.149	.8700	.0669	.0769	14.947	13.004	6.338	82.422	14
15	1.161	.8613	.0621	.0721	16.097	13.865	6.814	94.481	15
16	1.173	.8528	.0579	.0679	17.258	14.718	7.289	107.273	16
17	1.184	.8444	.0543	.0643	18.430	15.562	7.761	120.783	17
18	1.196	.8360	.0510	.0610	19.615	16.398	8.232	134.995	18
19	1.208	.8277	.0481	.0581	20.811	17.226	8.702	149.895	19
20	1.220	.8195	.0454	.0554	22.019	18.046	9.169	165.465	20
21	1.232	.8114	.0430	.0530	23.239	18.857	9.635	181.694	21
22	1.245	.8034	.0409	.0509	24.472	19.660	10.100	198.565	22
23	1.257	.7954	.0389	.0489	25.716	20.456	10.563	216.065	23
24	1.270	.7876	.0371	.0471	26.973	21.243	11.024	234.179	24
25	1.282	.7798	.0354	.0454	28.243	22.023	11.483	252.892	25
26	1.295	.7720	0339	.0439	29.526	22.795	11.941	272.195	26
27	1.308	.7644	.0324	.0424	30.821	23.560	12.397	292.069	27
28	1.321	.7568	.0311	.0411	32.129	24.316	12.852	312.504	28
29	1.335	.7493	.0299	.0399	33.450	25.066	13.304	333.486	29
30	1.348	.7419	.0287	.0387	34.785	25.808	13.756	355.001	30
36	1.431	.6989	.0232	.0332	43.077	30.107	16.428	494.620	36
40	1.489	.6717	.0205	.0305	48.886	32.835	18.178	596.854	40
48	1.612	.6203	.0163	.0263	61.223	37.974	21.598	820.144	48
50	1.645	.6080	.0155	.0255	64.463	39.196	22.436	879.417	50
52	1.678	.5961	.0148	.0248	67.769	40.394	23.269	939.916	52
60	1.817	.5504	.0122	.0222	81.670	44.955	26.533	1192.80	60
70	2.007	.4983	.00993	.0199	100.676	50.168	30.470	1528.64	70
72	2.047	.4885	.00955	.0196	104.710	51.150	31.239	1597.86	72
80	2.217	.4511	.00822	.0182	121.671	54.888	34.249	1879.87	80
84	2.307	.4335	.00765	.0177	130.672	56.648	35.717	2023.31	84
90	2.449	.4084	.00690	.0169	144.863	59.161	37.872	2240.56	90
96	2.599	.3847	.00625	.0163	159.927	61.528	39.973	2459.42	96
100	2.705	.3697	.00587	.0159	170.481	63.029	41.343	2605.77	100
104	2.815	.3553	.00551	.0155	181.464	64.471	42.688	2752.17	104
120	3.300	.3030	.00435	.0143	230.039	69.701	47.835	3334.11	120
240	10.893	.0918	.00101	.0110	989.254	90.819	75.739	6878.59	240
360	35.950	.0278	.00029	.0103	3495.0	97.218	89.699	8720.43	360
480	118.648	.00843	.00008	.0101	11,764.8	99.157	95.920	9511.15	480

Compound interest factors

$1\frac{1}{2}\%$ $\quad\quad\quad\quad\quad\quad\quad\quad\quad\quad\quad\quad\quad\quad\quad\quad\quad\quad\quad\quad\quad\quad\quad\quad\quad\quad\quad\quad\quad\quad\quad\quad\quad\quad\quad\quad\quad$ $1\frac{1}{2}\%$

	Single Payment		Uniform Payment Series				Uniform Gradient		
	Compound Amount Factor	Present Worth Factor	Sinking Fund Factor	Capital Recovery Factor	Compound Amount Factor	Present Worth Factor	Gradient Uniform Series	Gradient Present Worth	
n	Find F Given P F/P	Find P Given F P/F	Find A Given F A/F	Find A Given P A/P	Find F Given A F/A	Find P Given A P/A	Find A Given G A/G	Find P Given G P/G	n
1	1.015	.9852	1.0000	1.0150	1.000	0.985	0	0	1
2	1.030	.9707	.4963	.5113	2.015	1.956	0.496	0.970	2
3	1.046	.9563	.3284	.3434	3.045	2.912	0.990	2.883	3
4	1.061	.9422	.2444	.2594	4.091	3.854	1.481	5.709	4
5	1.077	.9283	.1941	.2091	5.152	4.783	1.970	9.422	5
6	1.093	.9145	.1605	.1755	6.230	5.697	2.456	13.994	6
7	1.110	.9010	.1366	.1516	7.323	6.598	2.940	19.400	7
8	1.126	.8877	.1186	.1336	8.433	7.486	3.422	25.614	8
9	1.143	.8746	.1046	.1196	9.559	8.360	3.901	32.610	9
10	1.161	.8617	.0934	.1084	10.703	9.222	4.377	40.365	10
11	1.178	.8489	.0843	.0993	11.863	10.071	4.851	48.855	11
12	1.196	.8364	.0767	.0917	13.041	10.907	5.322	58.054	12
13	1.214	.8240	.0702	.0852	14.237	11.731	5.791	67.943	13
14	1.232	.8118	.0647	.0797	15.450	12.543	6.258	78.496	14
15	1.250	.7999	.0599	.0749	16.682	13.343	6.722	89.694	15
16	1.269	.7880	.0558	.0708	17.932	14.131	7.184	101.514	16
17	1.288	.7764	.0521	.0671	19.201	14.908	7.643	113.937	17
18	1.307	.7649	.0488	.0638	20.489	15.673	8.100	126.940	18
19	1.327	.7536	.0459	.0609	21.797	16.426	8.554	140.505	19
20	1.347	.7425	.0432	.0582	23.124	17.169	9.005	154.611	20
21	1.367	.7315	.0409	.0559	24.470	17.900	9.455	169.241	21
22	1.388	.7207	.0387	.0537	25.837	18.621	9.902	184.375	22
23	1.408	.7100	.0367	.0517	27.225	19.331	10.346	199.996	23
24	1.430	.6995	.0349	.0499	28.633	20.030	10.788	216.085	24
25	1.451	.6892	.0333	.0483	30.063	20.720	11.227	232.626	25
26	1.473	.6790	.0317	.0467	31.514	21.399	11.664	249.601	26
27	1.495	.6690	.0303	.0453	32.987	22.068	12.099	266.995	27
28	1.517	.6591	.0290	.0440	34.481	22.727	12.531	284.790	28
29	1.540	.6494	.0278	.0428	35.999	23.376	12.961	302.972	29
30	1.563	.6398	.0266	.0416	37.539	24.016	13.388	321.525	30
36	1.709	.5851	.0212	.0362	47.276	27.661	15.901	439.823	36
40	1.814	.5513	.0184	.0334	54.268	29.916	17.528	524.349	40
48	2.043	.4894	.0144	.0294	69.565	34.042	20.666	703.537	48
50	2.105	.4750	.0136	.0286	73.682	35.000	21.428	749.955	50
52	2.169	.4611	.0128	.0278	77.925	35.929	22.179	796.868	52
60	2.443	.4093	.0104	.0254	96.214	39.380	25.093	988.157	60
70	2.835	.3527	.00817	.0232	122.363	43.155	28.529	1231.15	70
72	2.921	.3423	.00781	.0228	128.076	43.845	29.189	1279.78	72
80	3.291	.3039	.00655	.0215	152.710	46.407	31.742	1473.06	80
84	3.493	.2863	.00602	.0210	166.172	47.579	32.967	1568.50	84
90	3.819	.2619	.00532	.0203	187.929	49.210	34.740	1709.53	90
96	4.176	.2395	.00472	.0197	211.719	50.702	36.438	1847.46	96
100	4.432	.2256	.00437	.0194	228.802	51.625	37.529	1937.43	100
104	4.704	.2126	.00405	.0190	246.932	52.494	38.589	2025.69	104
120	5.969	.1675	.00302	.0180	331.286	55.498	42.518	2359.69	120
240	35.632	.0281	.00043	.0154	2308.8	64.796	59.737	3870.68	240
360	212.700	.00470	.00007	.0151	14,113.3	66.353	64.966	4310.71	360
480	1269.7	.00079	.00001	.0150	84,577.8	66.614	66.288	4415.74	480

Compound interest factors

2%   **2%**

	Single Payment		Uniform Payment Series				Uniform Gradient		
	Compound Amount Factor	Present Worth Factor	Sinking Fund Factor	Capital Recovery Factor	Compound Amount Factor	Present Worth Factor	Gradient Uniform Series	Gradient Present Worth	
n	Find F Given P F/P	Find P Given F P/F	Find A Given F A/F	Find A Given P A/P	Find F Given A F/A	Find P Given A P/A	Find A Given G A/G	Find P Given G P/G	n
1	1.020	.9804	1.0000	1.0200	1.000	0.980	0	0	1
2	1.040	.9612	.4951	.5151	2.020	1.942	0.495	0.961	2
3	1.061	.9423	.3268	.3468	3.060	2.884	0.987	2.846	3
4	1.082	.9238	.2426	.2626	4.122	3.808	1.475	5.617	4
5	1.104	.9057	.1922	.2122	5.204	4.713	1.960	9.240	5
6	1.126	.8880	.1585	.1785	6.308	5.601	2.442	13.679	6
7	1.149	.8706	.1345	.1545	7.434	6.472	2.921	18.903	7
8	1.172	.8535	.1165	.1365	8.583	7.325	3.396	24.877	8
9	1.195	.8368	.1025	.1225	9.755	8.162	3.868	31.571	9
10	1.219	.8203	.0913	.1113	10.950	8.983	4.337	38.954	10
11	1.243	.8043	.0822	.1022	12.169	9.787	4.802	46.996	11
12	1.268	.7885	.0746	.0946	13.412	10.575	5.264	55.669	12
13	1.294	.7730	.0681	.0881	14.680	11.348	5.723	64.946	13
14	1.319	.7579	.0626	.0826	15.974	12.106	6.178	74.798	14
15	1.346	.7430	.0578	.0778	17.293	12.849	6.631	85.200	15
16	1.373	.7284	.0537	.0737	18.639	13.578	7.080	96.127	16
17	1.400	.7142	.0500	.0700	20.012	14.292	7.526	107.553	17
18	1.428	.7002	.0467	.0667	21.412	14.992	7.968	119.456	18
19	1.457	.6864	.0438	.0638	22.840	15.678	8.407	131.812	19
20	1.486	.6730	.0412	.0612	24.297	16.351	8.843	144.598	20
21	1.516	.6598	.0388	.0588	25.783	17.011	9.276	157.793	21
22	1.546	.6468	.0366	.0566	27.299	17.658	9.705	171.377	22
23	1.577	.6342	.0347	.0547	28.845	18.292	10.132	185.328	23
24	1.608	.6217	.0329	.0529	30.422	18.914	10.555	199.628	24
25	1.641	.6095	.0312	.0512	32.030	19.523	10.974	214.256	25
26	1.673	.5976	.0297	.0497	33.671	20.121	11.391	229.169	26
27	1.707	.5859	.0283	.0483	35.344	20.707	11.804	244.428	27
28	1.741	.5744	.0270	.0470	37.051	21.281	12.214	259.936	28
29	1.776	.5631	.0258	.0458	38.792	21.844	12.621	275.703	29
30	1.811	.5521	.0247	.0447	40.568	22.396	13.025	291.713	30
36	2.040	.4902	.0192	.0392	51.994	25.489	15.381	392.036	36
40	2.208	.4529	.0166	.0366	60.402	27.355	16.888	461.989	40
48	2.587	.3865	.0126	.0326	79.353	30.673	19.755	605.961	48
50	2.692	.3715	.0118	.0318	84.579	31.424	20.442	642.355	50
52	2.800	.3571	.0111	.0311	90.016	32.145	21.116	678.779	52
60	3.281	.3048	.00877	.0288	114.051	34.761	23.696	823.692	60
70	4.000	.2500	.00667	.0267	149.977	37.499	26.663	999.829	70
72	4.161	.2403	.00633	.0263	158.056	37.984	27.223	1034.050	72
80	4.875	.2051	.00516	.0252	193.771	39.744	29.357	1166.781	80
84	5.277	.1895	.00468	.0247	213.865	40.525	30.361	1230.413	84
90	5.943	.1683	.00405	.0240	247.155	41.587	31.793	1322.164	90
96	6.693	.1494	.00351	.0235	284.645	42.529	33.137	1409.291	96
100	7.245	.1380	.00320	.0232	312.230	43.098	33.986	1464.747	100
104	7.842	.1275	.00292	.0229	342.090	43.624	34.799	1518.082	104
120	10.765	.0929	.00205	.0220	488.255	45.355	37.711	1710.411	120
240	115.887	.00863	.00017	.0202	5744.4	49.569	47.911	2374.878	240
360	1 247.5	.00080	.00002	.0200	62,326.8	49.960	49.711	2483.567	360
480	13 429.8	.00007		.0200	671,442.0	49.996	49.964	2498.027	480

Compound interest factors

4%   4%

	Single Payment		Uniform Payment Series				Uniform Gradient		
	Compound Amount Factor	Present Worth Factor	Sinking Fund Factor	Capital Recovery Factor	Compound Amount Factor	Present Worth Factor	Gradient Uniform Series	Gradient Present Worth	
n	Find F Given P F/P	Find P Given F P/F	Find A Given F A/F	Find A Given P A/P	Find F Given A F/A	Find P Given A P/A	Find A Given G A/G	Find P Given G P/G	n
1	1.040	.9615	1.0000	1.0400	1.000	0.962	0	0	1
2	1.082	.9246	.4902	.5302	2.040	1.886	0.490	0.925	2
3	1.125	.8890	.3203	.3603	3.122	2.775	0.974	2.702	3
4	1.170	.8548	.2355	.2755	4.246	3.630	1.451	5.267	4
5	1.217	.8219	.1846	.2246	5.416	4.452	1.922	8.555	5
6	1.265	.7903	.1508	.1908	6.633	5.242	2.386	12.506	6
7	1.316	.7599	.1266	.1666	7.898	6.002	2.843	17.066	7
8	1.369	.7307	.1085	.1485	9.214	6.733	3.294	22.180	8
9	1.423	.7026	.0945	.1345	10.583	7.435	3.739	27.801	9
10	1.480	.6756	.0833	.1233	12.006	8.111	4.177	33.881	10
11	1.539	.6496	.0741	.1141	13.486	8.760	4.609	40.377	11
12	1.601	.6246	.0666	.1066	15.026	9.385	5.034	47.248	12
13	1.665	.6006	.0601	.1001	16.627	9.986	5.453	54.454	13
14	1.732	.5775	.0547	.0947	18.292	10.563	5.866	61.962	14
15	1.801	.5553	.0499	.0899	20.024	11.118	6.272	69.735	15
16	1.873	.5339	.0458	.0858	21.825	11.652	6.672	77.744	16
17	1.948	.5134	.0422	.0822	23.697	12.166	7.066	85.958	17
18	2.029	.4936	.0390	.0790	25.645	12.659	7.453	94.350	18
19	2.107	.4746	.0361	.0761	27.671	13.134	7.834	102.893	19
20	2.191	.4564	.0336	.0736	29.778	13.590	8.209	111.564	20
21	2.279	.4388	.0313	.0713	31.969	14.029	8.578	120.341	21
22	2.370	.4220	.0292	.0692	34.248	14.451	8.941	129.202	22
23	2.465	.4057	.0273	.0673	36.618	14.857	9.297	138.128	23
24	2.563	.3901	.0256	.0656	39.083	15.247	9.648	147.101	24
25	2.666	.3751	.0240	.0640	41.646	15.622	9.993	156.104	25
26	2.772	.3607	.0226	.0626	44.312	15.983	10.331	165.121	26
27	2.883	.3468	.0212	.0612	47.084	16.330	10.664	174.138	27
28	2.999	.3335	.0200	.0600	49.968	16.663	10.991	183.142	28
29	3.119	.3207	.0189	.0589	52.966	16.984	11.312	192.120	29
30	3.243	.3083	.0178	.0578	56.085	17.292	11.627	201.062	30
31	3.373	.2965	.0169	.0569	59.328	17.588	11.937	209.955	31
32	3.508	.2851	.0159	.0559	62.701	17.874	12.241	218.792	32
33	3.648	.2741	.0151	.0551	66.209	18.148	12.540	227.563	33
34	3.794	.2636	.0143	.0543	69.858	18.411	12.832	236.260	34
35	3.946	.2534	.0136	.0536	73.652	18.665	13.120	244.876	35
40	4.801	.2083	.0105	.0505	95.025	19.793	14.476	286.530	40
45	5.841	.1712	.00826	.0483	121.029	20.720	15.705	325.402	45
50	7.107	.1407	.00655	.0466	152.667	21.482	16.812	361.163	50
55	8.646	.1157	.00523	.0452	191.159	22.109	17.807	393.689	55
60	10.520	.0951	.00420	.0442	237.990	22.623	18.697	422.996	60
65	12.799	.0781	.00339	.0434	294.968	23.047	19.491	449.201	65
70	15.572	.0642	.00275	.0427	364.290	23.395	20.196	472.479	70
75	18.945	.0528	.00223	.0422	448.630	23.680	20.821	493.041	75
80	23.050	.0434	.00181	.0418	551.244	23.915	21.372	511.116	80
85	28.044	.0357	.00148	.0415	676.089	24.109	21.857	526.938	85
90	34.119	.0293	.00121	.0412	827.981	24.267	22.283	540.737	90
95	41.511	.0241	.00099	.0410	1012.8	24.398	22.655	552.730	95
100	50.505	.0198	.00081	.0408	1237.6	24.505	22.980	563.125	100

Appendix A Engineering Economics

Compound interest factors

6%   **6%**

	Single Payment		Uniform Payment Series				Uniform Gradient		
	Compound Amount Factor	Present Worth Factor	Sinking Fund Factor	Capital Recovery Factor	Compound Amount Factor	Present Worth Factor	Gradient Uniform Series	Gradient Present Worth	
	Find F Given P F/P	Find P Given F P/F	Find A Given F A/F	Find A Given P A/P	Find F Given A F/A	Find P Given A P/A	Find A Given G A/G	Find P Given G P/G	
n									n
1	1.060	.943	1.0000	1.0600	1.000	0.943	0	0	1
2	1.124	.8900	.4854	.5454	2.060	1.833	0.485	0.890	2
3	1.191	.8396	.3141	.3741	3.184	2.673	0.961	2.569	3
4	1.262	.7921	.2286	.2886	4.375	3.465	1.427	4.945	4
5	1.338	.7473	.1774	.2374	5.637	4.212	1.884	7.934	5
6	1.419	.7050	.1434	.2034	6.975	4.917	2.330	11.459	6
7	1.504	.6651	.1191	.1791	8.394	5.582	2.768	15.450	7
8	1.594	.6274	.1010	.1610	9.897	6.210	3.195	19.841	8
9	1.689	.5919	.0870	.1470	11.491	6.802	3.613	24.577	9
10	1.791	.5584	.0759	.1359	13.181	7.360	4.022	29.602	10
11	1.898	.5268	.0668	.1268	14.972	7.887	4.421	34.870	11
12	2.012	.4970	.0593	.1193	16.870	8.384	4.811	40.337	12
13	2.133	.4688	.0530	.1130	18.882	8.853	5.192	45.963	13
14	2.261	.4423	.0476	.1076	21.015	9.295	5.564	51.713	14
15	2.397	.4173	.0430	.1030	23.276	9.712	5.926	57.554	15
16	2.540	.3936	.0390	.0990	25.672	10.106	6.279	63.459	16
17	2.693	.3714	.0354	.0954	28.213	10.477	6.624	69.401	17
18	2.854	.3503	.0324	.0924	30.906	10.828	6.960	75.357	18
19	3.026	.3305	.0296	.0896	33.760	11.158	7.287	81.306	19
20	3.207	.3118	.0272	.0872	36.786	11.470	7.605	87.230	20
21	3.400	.2942	.0250	.0850	39.993	11.764	7.915	93.113	21
22	3.604	.2775	.0230	.0830	43.392	12.042	8.217	98.941	22
23	3.820	.2618	.0213	.0813	46.996	12.303	8.510	104.700	23
24	4.049	.2470	.0197	.0797	50.815	12.550	8.795	110.381	24
25	4.292	.2330	.0182	.0782	54.864	12.783	9.072	115.973	25
26	4.549	.2198	.0169	.0769	59.156	13.003	9.341	121.468	26
27	4.822	.2074	.0157	.0757	63.706	13.211	9.603	126.860	27
28	5.112	.1956	.0146	.0746	68.528	13.406	9.857	132.142	28
29	5.418	.1846	.0136	.0736	73.640	13.591	10.103	137.309	29
30	5.743	.1741	.0126	.0726	79.058	13.765	10.342	142.359	30
31	6.088	.1643	.0118	.0718	84.801	13.929	10.574	147.286	31
32	6.453	.1550	.0110	.0710	90.890	14.084	10.799	152.090	32
33	6.841	.1462	.0103	.0703	97.343	14.230	11.017	156.768	33
34	7.251	.1379	.00960	.0696	104.184	14.368	11.228	161.319	34
35	7.686	.1301	.00897	.0690	111.435	11.498	11.432	165.743	35
40	10.286	.0972	.00646	.0665	154.762	15.046	12.359	185.957	40
45	13.765	.0727	.00470	.0647	212.743	15.456	13.141	203.109	45
50	18.420	.0543	.00344	.0634	290.335	15.762	13.796	217.457	50
55	24.650	.0406	.00254	.0625	394.171	15.991	14.341	229.322	55
60	32.988	.0303	.00188	.0619	533.126	16.161	14.791	239.043	60
65	44.145	.0227	.00139	.0614	719.080	16.289	15.160	246.945	65
70	59.076	.0169	.00103	.0610	967.928	16.385	15.461	253.327	70
75	79.057	.0126	.00077	.0608	1300.9	16.456	15.706	258.453	75
80	105.796	.00945	.00057	.0606	1746.6	16.509	15.903	262.549	80
85	141.578	.00706	.00043	.0604	2343.0	16.549	16.062	265.810	85
90	189.464	.00528	.00032	.0603	3141.1	16.579	16.189	268.395	90
95	253.545	.00394	.00024	.0602	4209.1	16.601	16.290	270.437	95
100	339.300	.00295	.00018	.0602	5638.3	16.618	16.371	272.047	100

Compound interest factors

8%   8%

	Single Payment		Uniform Payment Series				Uniform Gradient		
	Compound Amount Factor	Present Worth Factor	Sinking Fund Factor	Capital Recovery Factor	Compound Amount Factor	Present Worth Factor	Gradient Uniform Series	Gradient Present Worth	
n	Find F Given P F/P	Find P Given F P/F	Find A Given F A/F	Find A Given P A/P	Find F Given A F/A	Find P Given A P/A	Find A Given G A/G	Find P Given G P/G	n
1	1.080	.9259	1.0000	1.0800	1.000	0.926	0	0	1
2	1.166	.8573	.4808	.5608	2.080	1.783	0.481	0.857	2
3	1.260	.7938	.3080	.3880	3.246	2.577	0.949	2.445	3
4	1.360	.7350	.2219	.3019	4.506	3.312	1.404	4.650	4
5	1.469	.6806	.1705	.2505	5.867	3.993	1.846	7.372	5
6	1.587	.6302	.1363	.2163	7.336	4.623	2.276	10.523	6
7	1.714	.5835	.1121	.1921	8.923	5.206	2.694	14.024	7
8	1.851	.5403	.0940	.1740	10.637	5.747	3.099	17.806	8
9	1.999	.5002	.0801	.1601	12.488	6.247	3.491	21.808	9
10	2.159	.4632	.0690	.1490	14.487	6.710	3.871	25.977	10
11	2.332	.4289	.0601	.1401	16.645	7.139	4.240	30.266	11
12	2.518	.3971	.0527	.1327	18.977	7.536	4.596	34.634	12
13	2.720	.3677	.0465	.1265	21.495	7.904	4.940	39.046	13
14	2.937	.3405	.0413	.1213	24.215	8.244	5.273	43.472	14
15	3.172	.3152	.0368	.1168	27.152	8.559	5.594	47.886	15
16	3.426	.2919	.0330	.1130	30.324	8.851	5.905	52.264	16
17	3.700	.2703	.0296	.1096	33.750	9.122	6.204	56.588	17
18	3.996	.2502	.0267	.1067	37.450	9.372	6.492	60.843	18
19	4.316	.2317	.0241	.1041	41.446	9.604	6.770	65.013	19
20	4.661	.2145	.0219	.1019	45.762	9.818	7.037	69.090	20
21	5.034	.1987	.0198	.0998	50.423	10.017	7.294	73.063	21
22	5.437	.1839	.0180	.0980	55.457	10.201	7.541	76.926	22
23	5.871	.1703	.0164	.0964	60.893	10.371	7.779	80.673	24
24	6.341	.1577	.0150	.0950	66.765	10.529	8.007	84.300	24
25	6.848	.1460	.0137	.0937	73.106	10.675	8.225	87.804	25
26	7.396	.1352	.0125	.0925	79.954	10.810	8.435	91.184	26
27	7.988	.1252	.0114	.0914	87.351	10.935	8.636	94.439	27
28	8.627	.1159	.0105	.0905	95.339	11.051	8.829	97.569	28
29	9.317	.1073	.00962	.0896	103.966	11.158	9.013	100.574	29
30	10.063	.0994	.00883	.0888	113.283	11.258	9.190	103.456	30
31	10.868	.0920	.00811	.0881	123.346	11.350	9.358	106.216	31
32	11.737	.0852	.00745	.0875	134.214	11.435	9.520	108.858	32
33	12.676	.0789	.00685	.0869	145.951	11.514	9.674	111.382	33
34	13.690	.0730	.00630	.0863	158.627	11.587	9.821	113.792	34
35	14.785	.0676	.00580	.0858	172.317	11.655	9.961	116.092	35
40	21.725	.0460	.00386	.0839	259.057	11.925	10.570	126.042	40
45	31.920	.0313	.00259	.0826	386.506	12.108	11.045	133.733	45
50	46.902	.0213	.00174	.0817	573.771	12.233	11.411	139.593	50
55	68.914	.0145	.00118	.0812	848.925	12.319	11.690	144.006	55
60	101.257	.00988	.00080	.0808	1253.2	12.377	11.902	147.300	60
65	148.780	.00672	.00054	.0805	1847.3	12.416	12.060	149.739	65
70	218.607	.00457	.00037	.0804	2720.1	12.443	12.178	151.533	70
75	321.205	.00311	.00025	.0802	4002.6	12.461	12.266	152.845	75
80	471.956	.00212	.00017	.0802	5887.0	12.474	12.330	153.800	80
85	693.458	.00144	.00012	.0801	8655.7	12.482	12.377	154.492	85
90	1018.9	.00098	.00008	.0801	12,724.0	12.488	12.412	154.993	90
95	1497.1	.00067	.00005	.0801	18,701.6	12.492	12.437	155.352	95
100	2199.8	.00045	.00004	.0800	27,484.6	12.494	12.455	155.611	100

Compound interest factors

10%   **10%**

	Single Payment		Uniform Payment Series				Uniform Gradient		
	Compound Amount Factor	Present Worth Factor	Sinking Fund Factor	Capital Recovery Factor	Compound Amount Factor	Present Worth Factor	Gradient Uniform Series	Gradient Present Worth	
n	Find F Given P F/P	Find P Given F P/F	Find A Given F A/F	Find A Given P A/P	Find F Given A F/A	Find P Given A P/A	Find A Given G A/G	Find P Given G P/G	n
1	1.100	.9091	1.0000	1.1000	1.000	0.909	0	0	1
2	1.210	.8264	.4762	.5762	2.100	1.736	0.476	0.826	2
3	1.331	.7513	.3021	.4021	3.310	2.487	0.937	2.329	3
4	1.464	.6830	.2155	.3155	4.641	3.170	1.381	4.378	4
5	1.611	.6209	.1638	.2638	6.105	3.791	1.810	6.862	5
6	1.772	.5645	.1296	.2296	7.716	4.355	2.224	9.684	6
7	1.949	.5132	.1054	.2054	9.487	4.868	2.622	12.763	7
8	2.144	.4665	.0874	.1874	11.436	5.335	3.004	16.029	8
9	2.358	.4241	.0736	.1736	13.579	5.759	3.372	19.421	9
10	2.594	.3855	.0627	.1627	15.937	6.145	3.725	22.891	10
11	2.853	.3505	.0540	.1540	18.531	6.495	4.064	26.396	11
12	3.138	.3186	.0468	.1468	21.384	6.814	4.388	29.901	12
13	3.452	.2897	.0408	.1408	24.523	7.103	4.699	33.377	13
14	3.797	.2633	.0357	.1357	27.975	7.367	4.996	36.801	14
15	4.177	.2394	.0315	.1315	31.772	7.606	5.279	40.152	15
16	4.595	.2176	.0278	.1278	35.950	7.824	5.549	43.416	16
17	5.054	.1978	.0247	.1247	40.545	8.022	5.807	46.582	17
18	5.560	.1799	.0219	.1219	45.599	8.201	6.053	49.640	18
19	6.116	.1635	.0195	.1195	51.159	8.365	6.286	52.583	19
20	6.728	.1486	.0175	.1175	57.275	8.514	6.508	55.407	20
21	7.400	.1351	.0156	.1156	64.003	8.649	6.719	58.110	21
22	8.140	.1228	.0140	.1140	71.403	8.772	6.919	60.689	22
23	8.954	.1117	.0126	.1126	79.543	8.883	7.108	63.146	24
24	9.850	.1015	.0113	.1113	88.497	8.985	7.288	65.481	24
25	10.835	.0923	.0102	.1102	98.347	9.077	7.458	67.696	25
26	11.918	.0839	.00916	.1092	109.182	9.161	7.619	69.794	26
27	13.110	.0763	.00826	.1083	121.100	9.237	7.770	71.777	27
28	14.421	.0693	.00745	.1075	134.210	9.307	7.914	73.650	28
29	15.863	.0630	.00673	.1067	148.631	9.370	8.049	75.415	29
30	17.449	.0573	.00608	.1061	164.494	9.427	8.176	77.077	30
31	19.194	.0521	.00550	.1055	181.944	9.479	8.296	78.640	31
32	21.114	.0474	.00497	.1050	201.138	9.526	8.409	80.108	32
33	23.225	.0431	.00450	.1045	222.252	9.569	8.515	81.486	33
34	25.548	.0391	.00407	.1041	245.477	9.609	8.615	82.777	34
35	28.102	.0356	.00369	.1037	271.025	9.644	8.709	83.987	35
40	45.259	.0221	.00226	.1023	442.593	9.779	9.096	88.953	40
45	72.891	.0137	.00139	.1014	718.905	9.863	9.374	92.454	45
50	117.391	.00852	.00086	.1009	1163.9	9.915	9.570	94.889	50
55	189.059	.00529	.00053	.1005	1880.6	9.947	9.708	96.562	55
60	304.482	.00328	.00033	.1003	3034.8	9.967	9.802	97.701	60
65	490.371	.00204	.00020	.1002	4893.7	9.980	9.867	98.471	65
70	789.748	.00127	.00013	.1001	7887.5	9.987	9.911	98.987	70
75	1271.9	.00079	.00008	.1001	12,709.0	9.992	9.941	99.332	75
80	2048.4	.00049	.00005	.1000	20,474.0	9.995	9.961	99.561	80
85	3229.0	.00030	.00003	.1000	32,979.7	9.997	9.974	99.712	85
90	5313.0	.00019	.00002	.1000	53,120.3	9.998	9.983	99.812	90
95	8556.7	.00012	.00001	.1000	85,556.9	9.999	9.989	99.877	95
100	13,780.6	.00007	.00001	.1000	137,796.3	9.999	9.993	99.920	100

Compound interest factors

12%   12%

	Single Payment		Uniform Payment Series				Uniform Gradient		
	Compound Amount Factor	Present Worth Factor	Sinking Fund Factor	Capital Recovery Factor	Compound Amount Factor	Present Worth Factor	Gradient Uniform Series	Gradient Present Worth	
	Find F Given P F/P	Find P Given F P/F	Find A Given F A/F	Find A Given P A/P	Find F Given A F/A	Find P Given A P/A	Find A Given G A/G	Find P Given G P/G	
n									n
1	1.120	.8929	1.0000	1.1200	1.000	0.893	0	0	1
2	1.254	.7972	.4717	.5917	2.120	1.690	0.472	0.797	2
3	1.405	.7118	.2963	.4163	3.374	2.402	0.925	2.221	3
4	1.574	.6355	.2092	.3292	4.779	3.037	1.359	4.127	4
5	1.762	.5674	.1574	.2774	6.353	3.605	1.775	6.397	5
6	1.974	.5066	.1232	.2432	8.115	4.111	2.172	8.930	6
7	2.211	.4523	.0991	.2191	10.089	4.564	2.551	11.644	7
8	2.476	.4039	.0813	.2013	12.300	4.968	2.913	14.471	8
9	2.773	.3606	.0677	.1877	14.776	5.328	3.257	17.356	9
10	3.106	.3220	.0570	.1770	17.549	5.650	3.585	20.254	10
11	3.479	.2875	.0484	.1684	20.655	5.938	3.895	23.129	11
12	3.896	.2567	.0414	.1614	24.133	6.194	4.190	25.952	12
13	4.363	.2292	.0357	.1557	28.029	6.424	4.468	28.702	13
14	4.887	.2046	.0309	.1509	32.393	6.628	4.732	31.362	14
15	5.474	.1827	.0268	.1468	37.280	6.811	4.980	33.920	15
16	6.130	.1631	.0234	.1434	42.753	6.974	5.215	36.367	16
17	6.866	.1456	.0205	.1405	48.884	7.120	5.435	38.697	17
18	7.690	.1300	.0179	.1379	55.750	7.250	5.643	40.908	18
19	8.613	.1161	.0158	.1358	63.440	7.366	5.838	42.998	19
20	9.646	.1037	.0139	.1339	72.052	7.469	6.020	44.968	20
21	10.804	.0926	.0122	.1322	81.699	7.562	6.191	46.819	21
22	12.100	.0826	.0108	.1308	92.503	7.645	6.351	48.554	22
23	13.552	.0738	.00956	.1296	104.603	7.718	6.501	50.178	24
24	15.179	.0659	.00846	.1285	118.155	7.784	6.641	51.693	24
25	17.000	.0588	.00750	.1275	133.334	7.843	6.771	53.105	25
26	19.040	.0525	.00665	.1267	150.334	7.896	6.892	54.418	26
27	21.325	.0469	.00590	.1259	169.374	7.943	7.005	55.637	27
28	23.884	.0419	.00524	.1252	190.699	7.984	7.110	56.767	28
29	26.750	.0374	.00466	.1247	214.583	8.022	7.207	57.814	29
30	29.960	.0334	.00414	.1241	241.333	8.055	7.297	58.782	30
31	33.555	.0298	.00369	.1237	271.293	8.085	7.381	59.676	31
32	37.582	.0266	.00328	.1233	304.848	8.112	7.459	60.501	32
33	42.092	.0238	.00292	.1229	342.429	8.135	7.530	61.261	33
34	47.143	.0212	.00260	.1226	384.521	8.157	7.596	61.961	34
35	52.800	.0189	.00232	.1223	431.663	8.176	7.658	62.605	35
40	93.051	.0107	.00130	.1213	767.091	8.244	7.899	65.116	40
45	163.988	.00610	.00074	.1207	1358.2	8.283	8.057	66.734	45
50	289.002	.00346	.00042	.1204	2400.0	8.304	8.160	67.762	50
55	509.321	.00196	.00024	.1202	4236.0	8.317	8.225	68.408	55
60	897.597	.00111	.00013	.1201	7471.6	8.324	8.266	68.810	60
65	1581.9	.00063	.00008	.1201	13,173.9	8.328	8.292	69.058	65
70	2787.8	.00036	.00004	.1200	23,223.3	8.330	8.308	69.210	70
75	4913.1	.00020	.00002	.1200	40,933.8	8.332	8.318	69.303	75
80	8658.5	.00012	.00001	.1200	72,145.7	8.332	8.324	69.359	80
85	15,259.2	.00007	.00001	.1200	127,151.7	8.333	8.328	69.393	85
90	26,891.9	.00004		.1200	224,091.1	8.333	8.330	69.414	90
95	47,392.8	.00002		.1200	394,931.4	8.333	8.331	69.426	95
100	83,522.3	.00001		.1200	696,010.5	8.333	8.332	69.434	100

Compound interest factors

18%

	Single Payment		Uniform Payment Series				Uniform Gradient		
	Compound Amount Factor	Present Worth Factor	Sinking Fund Factor	Capital Recovery Factor	Compound Amount Factor	Present Worth Factor	Gradient Uniform Series	Gradient Present Worth	
n	Find F Given P F/P	Find P Given F P/F	Find A Given F A/F	Find A Given P A/P	Find F Given A F/A	Find P Given A P/A	Find A Given G A/G	Find P Given G P/G	n
1	1.180	.8475	1.0000	1.1800	1.000	0.847	0	0	1
2	1.392	.7182	.4587	.6387	2.180	1.566	0.459	0.718	2
3	1.643	.6086	.2799	.4599	3.572	2.174	0.890	1.935	3
4	1.939	.5158	.1917	.3717	5.215	2.690	1.295	3.483	4
5	2.288	.4371	.1398	.3198	7.154	3.127	1.673	5.231	5
6	2.700	.3704	.1059	.2859	9.442	3.498	2.025	7.083	6
7	3.185	.3139	.0824	.2624	12.142	3.812	2.353	8.967	7
8	3.759	.2660	.0652	.2452	15.327	4.078	2.656	10.829	8
9	4.435	.2255	.0524	.2324	19.086	4.303	2.936	12.633	9
10	5.234	.1911	.0425	.2225	23.521	4.494	3.194	14.352	10
11	6.176	.1619	.0348	.2148	28.755	4.656	3.430	15.972	11
12	7.288	.1372	.0286	.2086	34.931	4.793	3.647	17.481	12
13	8.599	.1163	.0237	.2037	42.219	4.910	3.845	18.877	13
14	10.147	.0985	.0197	.1997	50.818	5.008	4.025	20.158	14
15	11.974	.0835	.0164	.1964	60.965	5.092	4.189	21.327	15
16	14.129	.0708	.0137	.1937	72.939	5.162	4.337	22.389	16
17	16.672	.0600	.0115	.1915	87.068	5.222	4.471	23.348	17
18	19.673	.0508	.00964	.1896	103.740	5.273	4.592	24.212	18
19	23.214	.0431	.00810	.1881	123.413	5.316	4.700	24.988	19
20	27.393	.0365	.00682	.1868	146.628	5.353	4.798	25.681	20
21	32.324	.0309	.00575	.1857	174.021	5.384	4.885	26.330	21
22	38.142	.0262	.00485	.1848	206.345	5.410	4.963	26.851	22
23	45.008	.0222	.00409	.1841	244.487	5.432	5.033	27.339	24
24	53.109	.0188	.00345	.1835	289.494	5.451	5.095	27.772	24
25	62.669	.0160	.00292	.1829	342.603	5.467	5.150	28.155	25
26	73.949	.0135	.00247	.1825	405.272	5.480	5.199	28.494	26
27	87.260	.0115	.00209	.1821	479.221	5.492	5.243	28.791	27
28	102.966	.00971	.00177	.1818	566.480	5.502	5.281	29.054	28
29	121.500	.00823	.00149	.1815	669.447	5.510	5.315	29.284	29
30	143.370	.00697	.00126	.1813	790.947	5.517	5.345	29.486	30
31	169.177	.00591	.00107	.1811	934.317	5.523	5.371	29.664	31
32	199.629	.00501	.00091	.1809	1103.5	5.528	5.394	29.819	32
33	235.562	.00425	.00077	.1808	1303.1	5.532	5.415	29.955	33
34	277.963	.00360	.00065	.1806	1538.7	5.536	5.433	30.074	34
35	327.997	.00305	.00055	.1806	1816.6	5.539	5.449	30.177	35
40	750.377	.00133	.00024	.1802	4163.2	5.548	5.502	30.527	40
45	1716.7	.00058	.00010	.1801	9531.6	5.552	5.529	30.701	45
50	3927.3	.00025	.00005	.1800	21,813.0	5.554	5.543	30.786	50
55	8984.8	.00011	.00002	.1800	49,910.1	5.555	5.549	30.827	55
60	20,555.1	.00005	.00001	.1800	114,189.4	5.555	5.553	30.846	60
65	47,025.1	.00002		.1800	261,244.7	5.555	5.554	30.856	65
70	107,581.9	.00001		.1800	597,671.7	5.556	5.555	30.860	70
75	46,122.1				1,367,339.2	5.556	5.555	30.862	75
100	15,424,131.9				85,689,616.2	5.556	5.555	30.864	100